ORGANIZING COOLS THE PLANET

TOOLS AND REFLECTIONS TO NAVIGATE THE CLIMATE CRISIS

HILARY MOORE & JOSHUA RUSSELL

PM Press PAMPHLET SERIES

0001: BECOMING THE MEDIA: A CRITICAL HISTORY OF CLAMOR MAGAZINE
By Jen Angel

0002: DARING TO STRUGGLE, FAILING TO WIN: THE RED ARMY FACTION'S 1977 CAMPAIGN OF DESPERATION
By J. Smith and André Moncourt

0003: MOVE INTO THE LIGHT: POSTSCRIPT TO A TURBULENT 2007
By the Turbulence Collective

0004: THE PRISON-INDUSTRIAL COMPLEX AND THE GLOBAL ECONOMY
By Eve Goldberg and Linda Evans

0005: ABOLISH RESTAURANTS: A WORKER'S CRITIQUE OF THE FOOD SERVICE INDUSTRY
By Prole

0006: SING FOR YOUR SUPPER: A DIY GUIDE TO PLAYING MUSIC, WRITING SONGS, AND BOOKING YOUR OWN GIGS
By David Rovics

0007: PRISON ROUND TRIP
By Klaus Viehmann

0008: SELF-DEFENSE FOR RADICALS: A TO Z GUIDE FOR SUBVERSIVE STRUGGLE
By Mickey Z.

0009: SOLIDARITY UNIONISM AT STARBUCKS
By Staughton Lynd and Daniel Gross

0010: COINTELSHOW: A PATRIOT ACT
By L.M. Bogad

0011: ORGANIZING COOLS THE PLANET: TOOLS AND REFLECTIONS TO NAVIGATE THE CLIMATE CRISIS
By Hilary Moore and Joshua Kahn Russell

0012: "VENCEREMOS": VÍCTOR JARA AND THE NEW CHILEAN SONG MOVEMENT
By Gabriel San Román

0013: ON COMMUNITY CIVIL DISOBEDIENCE IN THE NAME OF SUSTAINABILITY: THE COMMUNITY RIGHTS MOVEMENT IN THE UNITED STATES
By the Community Environmental Legal Defense Fund with an introduction by Thomas Linzey

0014: THAT PRECIOUS STRAND OF JEWISHNESS THAT CHALLENGES AUTHORITY
By Leon Rosselson

0015: DIVIDE AND CONQUER OR DIVIDE AND SUBDIVIDE? HOW NOT TO REFIGHT THE FIRST INTERNATIONAL
By Mark Leier

0016: HEART X-RAYS: A MODERN EPIC POEM
By G.H. Mosson and Marcus Colasurdo

0017: THE YOUNG C.L.R. JAMES: A GRAPHIC NOVELETTE
By Milton Knight • Edited by Lawrence Ware and Paul Buhle

0018: EVENTS AND VICTIMS
By Bartolomeo Vanzetti • Edited by Jon Curley

PM Press PAMPHLET SERIES No. 0011
ORGANIZING COOLS THE PLANET: TOOLS AND REFLECTIONS TO NAVIGATE THE CLIMATE CRISIS
By Joshua Kahn and Hilary Moore
ISBN: 978-1-60486-443-4

PM Press
PO Box 23912
Oakland, CA 94623
www.pmpress.org

Cover Art: Nate Vieland from the Beehive Design Collective
Interior Art: Joshua Kahn and Beatriz Carmen Mendoza. Image on page 24 by Ricardo Levins Morales; Image on page 48 inspired by a cartoon by S. Gross
Layout: Jonathan Rowland

TABLE OF CONTENTS

> One of the great liabilities of history is that all too many people fail to remain awake through great periods of social change. Every society has its protectors of the status quo and its fraternities of the indifferent who are notorious for sleeping through revolutions. But today our very survival depends on our ability to stay awake, to adjust to new ideas, to remain vigilant and to face the challenge of change.
>
> —Dr. Martin Luther King Jr.

INTRO

ONE DAY THE TWO OF US HAD A CONVERSATION

Hey Hilary, it's good to see you! How are you?

Hi Josh, I'm doing pretty good, but I've just been feeling really challenged, ya know?

Yeah, by what?

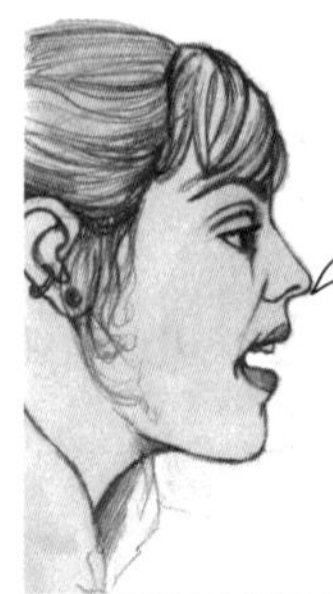

Well, it's kind of hard to put into words, but I'm craving more of a dialogue with critical questions about how we can help build a movement for Climate Justice. I'm thinking about the work I do with Mobilization for Climate Justice West, and about how those of us who aren't from "frontline" communities* can deeply engage in this movement. The more I dig for answers the more I land on more questions.

I've been feeling the same way. I just got back from our Ruckus Society training camp for Eco-Justice activists, and I keep thinking about how the scale of the ecological crisis can be overwhelming. Around the campfire, I'd ask other organizers how they really feel about their work, and there's a lot of despair underneath it all. What happens when our theories of change aren't producing change at the rate or scale it needs to happen?

*For a fuller explanation and definition of "frontline" community, see the section titled "The point of this booklet" on page 11.

You know what? We need to rethink the questions we should be asking.

Definitely. What if we wrote down some of this stuff, like our stories, experiences, missteps, and successes, and reflected on them? It would be a great opportunity to bounce ideas off of other organizers and start a collaborative conversation. Do you think it would be useful to other people in similar roles?

Totally! We both know so many people who are eager for more dialogue. You know, the way we present ourselves is key: we're not pretending to be experts, but I do think we've come across some really important questions and situations that need to be shared. The Climate Justice movement is developing so quickly that I feel a lot of really important stuff goes unnoticed. I've been wanting to do a zine for groups in our position about accountability and stuff, so this fits in great.

Maybe if we write a booklet, it will help start conversations on some of these elephants in the room, people can share their experiences too, and we can start to navigate our way through a pretty confusing political landscape here in North America. We could have our booklet model the kind of accountability we want to practice in our organizing, like offering the questions we're thinking about and getting all kinds of feedback and direction from frontline communities and organizers.

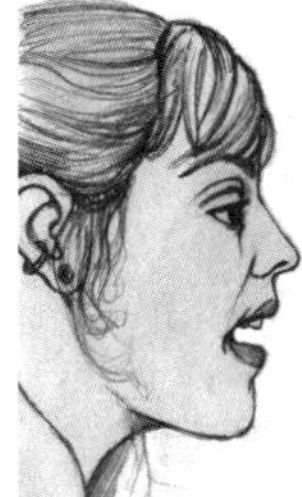
Yes! Our process can mirror our organizing practice, and we can use it to examine some of the unique challenges in working for Climate Justice when we're doing solidarity work. Hopefully the project can grow—as others reflect and add their experiences, we can expand it more.

...and here we are now. We generated a lot more material than we could fit in this booklet. We're excited to get the ball rolling. We're offering this as a starting point. Before we jump in, here are some clarifications about what we are, and are not, trying to do:

THIS BOOKLET IS

- A starting point. We hope these ideas will be quickly deepened or replaced by our peers as we expand and extend this conversation.
- An articulation of a political framework (Climate Justice) to understand some of the challenges we face and respond to them. It isn't static. It isn't the *only* useful framework in addressing climate change, either. But while the parameters and definitions of Climate Justice are in flux, we are articulating the current framework we've been operating with at the moment.
- An organizing tool that groups and individuals can use to navigate the North American Climate Justice *movement* led by impacted communities (primarily low- income communities, communities of color, and Indigenous peoples).
- An honest reflection on the political frameworks, ideas, and practices climate activists have used, and how they have fared in making change around us. How do our models look in real life? Are they working?
- A set of tools and insights that are reflective of our personal journeys, limited by our own experience and context.
- Intended for organizers who are having similar challenges to ours. We're writing for people like us.

THIS BOOKLET IS NOT

- Trying to tell anyone else how to organize their community or create some new standard of ethical purity or unachievable set of ideals that don't play out in the nitty-gritty real world we live in.
- A 101 breakdown of climate science or how climate change works.[1]
- A comprehensive international policy solution.[2]
- A deep engagement of the active debates around "what is climate justice?": its evolving definition, "who owns it," its misuses, etc.
- A set of universally applicable formulas, rules, or dogmas that can be applied in any context, handed down from "experts."

WHERE WE'RE COMING FROM

Many of the ideas, frameworks, and tools in this booklet come from organizations we have learned from in our journeys. This includes but is not limited to: Movement Generation Justice and Ecology Project, smartMeme Strategy & Training Project, Training for Change, the Ruckus Society, Beyond the Choir, Climate Justice Now!, Mobilization for Climate Justice West, Indigenous Environmental Network, and Rising Tide. We encourage all who find these tools useful to check out the work of these outstanding groups.

---

1 If you're reading this, we assume that you understand the basics of how climate change works, that it's human-made, and that our actions matter in shaping the course of the future.

2 For resources on international policy, see the Institute for Policy Studies site, www.ips-dc.org, and many of the organizations listed in the Resources section of this booklet.

While drafting this booklet, many people helped give us direction, feedback, edits, and ideas throughout an extensive feedback and accountability process.[3] While it's impossible to list everyone who has influenced our thinking, the people who helped directly with this project are: Maryam Adrangi, Mel Bazil, May Boeve, Patrick Bond, Doyle Canning, Gopal Dayeneni, Michael Dorsey, Carol Duong, Madeline Gardner, Jihan Gearon, Tom Goldtooth, Jamie Henn, Ben Holtzman, Cathy Kunkel, Sharon Lungo, Michelle Mascarenhas-Swan, Tadzio Muller, Ana Orozco, Payal Parekh, Scott Parkin, Diana Pei Wu, Carla Pérez, Anne Petermann, Ben Powless, Patrick Reinsborough, Rafter Sass, Levana Saxon, Emily Simons, Matt Smucker, David Solnit, Brian Tokar, Jessica Tovar, Kevin Van Meter, Dave Vasey, and Ian Vitteri.

3 For more on our methodology, see the end of this booklet.

CHAPTER ONE FIND YOUR FRONTLINE

> The challenge of modernity is to live without illusions and without becoming disillusioned.
>
> —Antonio Gramsci

THE WORLD AS WE SEE IT

Two days before the manuscript for this booklet was due, Japan had the largest earthquake in its history, pushing tsunami waves across the entire Pacific Ocean.[4] The earthquake shifted Japan's coastline eight feet and tilted the whole Earth's axis.[5] This year, the U.S. National Oceanic and Atmospheric Administration recorded the highest land temperatures in human history, yet again. Immediate effects include fifteen thousand heat deaths in Russia, accompanied by record wildfires devastating crops and skyrocketing prices of corn and wheat.[6] Record droughts have ravaged Pakistan. In Latin America, record rainfalls washed away entire mountainsides. While the two of us refuse to be paralyzed by the end-of-the-world-mongers, it is undeniable that we're living in exponential times.

But digging up all the carbon Mother Nature sequestered underground and burning it for energy didn't just *happen*. It was the result of an ever-expanding global economy. Turns out, it's impossible to have infinite growth on a finite planet. The basic logic in our economic system is bursting.

Just as the climate crisis is one *symptom* of our current economic system, we're facing other ecological crises too: fisheries are collapsing, species are disappearing faster than the last great extinction, drinkable water is vanishing... The climate crisis is just one expansive element of many overlapping crises associated with the collapse of the ecological systems that support life on this planet. In fact, we're living during a period of the most rapid transformation in human history on all facets of life: technological[7], political, cultural, ecological, economic...

Okay. That's a bit overwhelming. But it's clear that stoking our cultural fear of some "doomsday" in the future is not useful for building organizations, community, and mobilizing people to create a livable future; or for preventing the worst impacts of runaway climate chaos.

---

4 While there are studies that connect global warming with increased seismic activity, we include it here not because the Japanese earthquake was necessarily the direct result of climate change, but because it reminds us of the power of Mother Nature. As the death toll rises, it underscores the need for resilient communities in the face of increased disasters. As Japan now faces the threat of nuclear crisis, Naomi Klein reminds us that *no energy that can poison entire populations during disasters is "clean."* A turbulent future needs nontoxic energy.

5 www.cnn.com/2011/WORLD/asiapcf/03/12/japan.earthquake.tsunami.earth/index.html?on.cnn=1

6 www.bloomberg.com/news/2011-01-12/land-surface-temperatures-set-record-high-in-2010-noaa-says.html

7 Even the neurology of how human brains develop has changed with the advent of the internet, rapid communication, and smart phones. See www.beyondthechoir.org for analysis on neurology and social movements.

People often talk of climate change as a single apocalyptic event that may happen in the future; either humanity averts it, or we do not. The reality is that the impacts of climate change have been happening for quite a while to peoples across the planet and are now increasing in their number, severity, and location. In fact, climate destabilization accelerates and amplifies the other crises that marginalized people have been dealing with for generations.[8]

Therefore, we see our main challenge as finding ways to *navigate the multiple crises that are already changing our lives*. What we do next will determine the scale and the scope of our transition. When we think about it that way, having *hope* at the center of our work makes sense in the face of such overwhelming odds.

So if collapses are already here, and we see more and more coming, our question isn't whether we can "solve the climate crisis," but *how do we navigate change? Will there be justice on the other side*? That is our work together. That is where we find vision and inspiration.

Liberation struggles throughout history have always been urgent. They have always been life-or-death, but each have had their own timeline.[9] For example, if you are working to decolonize your country from a European occupier, you fight until you win. The ecological crisis we face has that dimension, *plus* a science-based timeline that we can't negotiate with. What we do in the next two years will determine the landscape for the next ten years, which will determine the landscape for the next one hundred years.[10]

No pressure.

Because of this timeline imposed on us by Nature (and revealed by science), social movements must ask *new questions* that we wouldn't be asking if we had all the time we wanted. Because ecological collapse is embedded in all aspects of life on this planet, we need to think about *scale* in new ways. Social movements need to make unlikely alliances *that we wouldn't otherwise make*. We therefore need a political compass with which to navigate these choices in a strategic and principled way.

This booklet is an effort to tune that compass for the task ahead of all of us. We hope it will be useful to you too, in locating yourself and your work in relation to the expanding North American Climate Justice movement.

The least likely future is one in which things stay the same. One way or another, the existing culture and economy will need to shift to meet the ecological disruptions it is causing.

> We may face a scorched and lifeless earth. But they're accountable to their shareholders first. That's how the world works.
>
> —Propagandhi

---

8 Such as access to food, clothing, housing, and education; freedom of migration and mobility; and freedom from war, racial profiling, or police brutality.

9 We can learn from recent global mass struggles that had political timelines. For example, in the 1980s under Reagan, the threat of nuclear war was perceived as likely and imminent. Successful mass movements for nuclear disarmament brought about deep and long-ranging human and ecological impacts.

10 Unfortunately, some groups have mistaken the urgency of our crisis as a reason to abandon a commitment to the long-haul effort of transformative social justice. For more on the "Emergency-Mode Trojan Horse" see page 21.

Some of those in power are squinting over these turbulent, unpredictable waters along with us, wondering how to make it across. Corporations are finding ways to fleece our crises, proposing schemes to make money from our disasters and calling them "climate solutions." Pentagon projections focus on new kinds of warfare in a resource-depleted landscape driven by climate chaos.[11] It's clear that the business-as-usual crowd *is* thinking about scale in this transition. Their main motivation is to keep the big "status quo" ship sailing through the storm, maintaining their own wealth and power as much as they can. The "status quo" is dynamic and adaptive.

Unlike the power-holders, when we gaze across the rough sea, *our* motivation is to create the most equitable lifeboats possible, so we can all live well through what will be a bumpy ride. *Our task is to navigate change differently than those in power.*

11 See *Alternative Futures and Army Force Planning: Implications for the Future Force Era* by the RAND Corporation for military projections, http://www.rand.org/pubs/monographs/MG219.html

THE POINT OF THIS BOOKLET

Find your frontline[12] — Align it with others'

Walk the street with us into history. Get off the sidewalk.

—Dolores Huerta

We're still wrestling with our own particular roles inside movements for social change in general, and understanding the framework of "climate justice" in particular. Unsatisfied with the notions of "solidarity" (though we also do solidarity work) or the identity of "ally" (though we also consider ourselves allies[13]) that used to offer us clarity, we searched for a way to think of our work that can holistically capture the nuance of our moment.

We are part of a community of practice that has been thinking through useful ways to understand these concepts, and we owe a lot to the ideas of our peers and mentors. The thinking in this section is drawn heavily from contributions from grassroots organizers involved in the 2010 U.S. Social Forum EcoJustice People's Movement Assembly[14] process, the "CJ in the USA: Root-Cause Remedies, Rights, Reparations, and Representation statement,"[15] the "Grassroots Organizing Cools the Planet" open letter[16] (inspired by La Via Campesina's slogan "Small Farmers Cool the Planet'), the work of Movement Generation,[17] Occidental Arts and Ecology Center, and others who have been making thoughtful contributions to how we articulate our challenge.

The fact is that climate change *does* affect everybody. We believe that in order to build a popular movement, we need to reach out to everyone we can to help them understand the ways that their own lives are impacted by the crisis, and that there are real ways to take action. Therefore we need to *start* by helping people name their own impact. But as organizers, we need to take a step back first, and understand the context we're operating in. The slogan "We're all in this together" is true but misleading: climate change certainly doesn't affect everybody *equally*.

Environmental Justice

Our framework starts by honoring the roots of the U.S. Environmental Justice (EJ) movement[18] and how it has contributed to Climate Justice. While we all have a stake in a livable planet, the insight that EJ offers the world is understanding *disproportionate*

---

12 The concept "Find Your Frontline" was coined by Movement Generation in 2010.

13 If you are unfamiliar with the concept of being an "ally" or solidarity, here is a great primer of the value of this role, adapted from Paul Kivel's book *Uprooting Racism: How White People Can Work for Social Justice*, http://www.paulkivel.com/articles/guidelinesforbeingstrongwhiteallies.pdf

14 See the EcoJustice PMA declaration, http://itsgettinghotinhere.org/2010/06/29/declaration-from-the-us-social-forums-ecojustice-people%E2%80%99s-movement-assembly/

15 http://www.climate-justice-now.org/cj-in-the-usa-root-cause-remedies-rights-reparations-and-representation/

16 http://www.grist.org/article/2010-10-23-open-letter-to-1-sky-from-the-grassroots

17 Gopal Dayaneni in particular was been pivotal to helping us work through these ideas.

18 The roots of the Environmental Justice movement are in confronting and removing the point-source pollution that ends up accumulating in the communities of Indigenous, low-income people, and people of color. For more resources on EJ see, http://www.ejnet.org/ej/.

impact.[19] The Environmental Justice sense of "impact" is how racism and poverty determine which communities choke on exhaust from incinerators and refineries, and which communities have their land razed and resources taken to power U.S. cities. Those of us who have the luxury of turning on our lights and not thinking about where that power came from have a lot of privilege at the expense of others. That privilege determines which communities *don't* suffer from skyrocketing rates of asthma or leukemia, rare cancers, and other manifestations of toxic dumping, spewing, and pumping.

Impact

When we move from Environmental Justice to Climate Justice, the ways we think about impact become much broader. We're not just focusing on root causes of point-source pollution or heavy metals in the water supply; we're taking a step back and looking at other parts of the broader climate crisis, too. From that vantage point, frontlines are *all around us*, and the "win" is transformation of the economy and our relationships.

Ways you are impacted might be...

- Your sister or brother is stationed in the ongoing occupation of Iraq or another war to secure fossil fuel resources.
- You live in California where two nuclear power plants are sited on fault lines, just like Japan.
- You are a U.S.-born child of undocumented immigrants in Phoenix, Arizona, facing deportation. You find there are some out-of-touch environmentalists using "resource consumption" and population growth as an argument in support of inhumane anti-immigration laws.
- Your family has worked in the Michigan auto industry for generations, but no one in your town can find those jobs since they've been outsourced to the global economy.
- You live on the coast of Maine and the fisheries that have sustained your family for generations are no longer viable.
- You are a union teacher in Wisconsin who just lost your collective bargaining rights under the pretence of "balancing the budget" in a recession driven by chronic insecurity.
- You live in a place like Las Vegas, which will have a watershed that is incapable of supporting its population in the foreseeable future.
- You live in the Gulf Coast and the oil spills have destroyed your family's ability to fish for a living.

As we can see, not all "impact" is equal or the same, but when we look at the economic and political roots of the climate crisis, we can all find ways we're affected.

Community

For now, let's define "community" loosely as a group of people. We'll zoom in to some of the different kinds of communities and the ways organizers relate to them on page 27.

19 Common examples of "EJ communities" are those suffering near coal mines, oil refineries, or uranium mines, and communities in areas prone to the droughts, famines, floods, and hurricanes that come with climate change who additionally face institutional neglect (like New Orleans after Hurricane Katrina).

Frontline

The reason we differentiate between "frontline" communities and "impacted" communities is an added layer of *action*. Frontline communities are directly impacted communities who have been able to collectively *name the ways they are burdened* and are *organizing for action together.*

Frontline Solutions

Organizing for action doesn't just mean stopping the bad stuff. We believe in building a movement that lifts up, and leads with, climate *solutions* from the frontlines. The communities most directly impacted on the frontlines are not only dealing with the brunt of the problem, but are also best equipped with the knowledge and skills to chart the way forward. For example, the cycle of globalized industrial extraction, production, consumption and waste, also produces chronic food insecurity. Over one billion people struggling to afford more than one meal a day are mostly small farming communities that know best how to feed people, in their own context. Such traditional farming techniques remain sustainable and viable when freed from the stranglehold of the global economy. This doesn't mean that frontline communities "have all the answers" or can offer a one-size-fits-all solution to a global crisis, but it does mean that when building viable solutions, we don't need to start from scratch. Humans have known how to live in balance with their environment for the majority of our history, and many still do today. Furthermore, because frontline communities are by definition *organized,* they are in a position to *apply* their knowledge to the problem.[20]

Tuning Our Political Compass:

- How can taking inspiration from frontline solutions inform your own solutions where you live?
- How do we think about the *scale* of our solutions to make a credible case for closing the door to geoengineering and other proposals that make our problem worse?[21]
- There are lots of community solutions everywhere you look, but often people don't see them. How can we connect them into a broader popular narrative that is accessible and compelling?

Learning from these communities helps all of us envision what genuine solutions look like in our own communities and gives us inspiration to build them. That's why there's room for everybody in this movement. We need everybody in this movement if we're going to navigate the transition. But it does mean that our roles are different

20 For example, forest-dwelling Indigenous communities that are negatively impacted by false solutions like offsetting schemes have been attending the United Nations Framework Convention on Climate Change to advocate *real* solutions based on their unique knowledge inherited from generations of forest stewardship.

21 The Etc. Group recently put out an excellent primer on geoengineering that you can download at http:// www.etcgroup.org/en/issues/geoengineering

depending on our impact. So what happens if I am not from an "EJ community" or a place that fits traditional notions of what a "frontline community" is? What's my role in this work?

Find Your Frontline

My role becomes clearer if I find my frontline. Figure out the material and systemic impact that climate change has on me and my community, name it, and *get organized around it.*

Maybe *that's* where I should take action—and maybe not. Everyone has a frontline, but not all frontlines are equally strategic.

Here's an example. Both of us writing this booklet have been relatively insulated from skyrocketing food prices or toxic emissions from the polluting industries that are screwing up our planet, and we can see that that very insulation has shifted burden further on the poor. But we also know that eventually water insecurity will hit us. In that case, that particular frontline may not feel so urgent that it's constraining our immediate choices today. Maybe the best way to serve our own communities—to intervene on our frontlines—is to throw down directly with other folks on *their* frontline.

The mantra "think globally, act locally" works for *individual action* but quickly breaks down when we're thinking about *systematic change,* since not all localities are equally relevant to the global economy. Instead, Climate Justice calls on us to "think structurally, act strategically" (though it may not be as catchy on a bumper sticker).[22] Therefore, if your frontline doesn't offer immediate action opportunities (or if they are not impactful at this moment), you need to align it with other frontlines.

Taking Action: Aligning Your Frontline with Others'

If "finding your frontline" is about your relation to *impact,* "aligning your frontline" is about the *relationship to our political moment.* It is fundamentally about *strategy.*

We can assess that some frontline fights are critical to all other struggles, and can help us all change the game. For example, coal is a key piece of the fossil fuel puzzle. Right now there is a critical fight happening in Appalachia around ending mountaintop removal coal mining. If it's won, many other struggles become a little bit easier.[23] The same can be said for the current union fights for collective bargaining rights in Wisconsin and Indiana.

Ask yourself or your group:

- What does our frontline have in common with another frontline that we want to align ourselves with?
- What is different about our relation to impact?
- What is our common ground politically? Where can we collaborate?
- How do our differences build barriers? How can we navigate and overcome these barriers?

---

22 Slogan inspired by Patrick Reinsborough.

23 Of course this is a simplification: the impact on other sectors or fossil fuel struggles also depends on *how* this fight is won.

A key piece of this strategic assessment is that it is difficult, if not impossible, to do alone. That's why "finding your frontline" starts with naming your impact and working in a *group*. With a group, you can assess your capacity and priorities much more effectively than an individual drifting in the stormy sea. Our political landscape is shifting, offering us different pressure points, where we have unique opportunities, or where our opponents are particularly vulnerable. Even though many local communities each have their own priorities and urgent fights they must focus on,[24] we all need common strategic frameworks to move together or collectively assess critical moments[25] that need national or international solidarity. That is why interlocking action is grounded in community organizing, but certainly not limited to it. There are many large-scale flashpoints which require other complimentary avenues for action.[26]

Solidarity Organizing

> If you have come to help me, then you are wasting your time. But if you have come because your liberation is bound up with mine, then let us work together.
>
> —Lila Watson

When you align your organizing with someone else's frontline, you're practicing a form of "solidarity." Solidarity organizing isn't one-directional. We don't practice solidarity just because we're ideologically committed to it; we practice solidarity because it's strategic. Instead of trying to motivate our peers through altruism, we help them understand that this is the way we can win. This helps us move beyond patterns of paternalism—those who have resources "helping" those who do not. It's when the solidarity activist is unrooted, disconnected from their own history or impact, that the worst patterns of appropriation, arrogance, or savior complexes rear their ugly heads. *You need to know who you are in order to work well with those different from you.* You will be effective to the degree that you understand how your frontline relates to others. The process of "alignment" is the painstaking work of organizing—taking into account strategy, power, privilege, access, impact, difference, similarity, trust—but it produces a movement in which we're not acting on behalf of one another; *we can take meaningful action in an interlocking way.*

---

24 Many directly impacted frontline communities do not have the luxury of choosing their fight. Their battlefields are chosen for them by circumstance and injustice.

25 In addition to the mobilizations in Wisconsin and Indiana around collective bargaining occurring at the time of this writing, other recent critical movement flashpoints have included fighting anti-immigrant laws passed in Arizona in the summer of 2009, or internationally at the Copenhagen COP15 UN Climate Negotiations. These moments often forecast the future or set the tone for the country as a whole.

26 This could include going after various points in the fossil fuel chain of destruction, corporate accountability, federal policy work, local institution building, national media spectacle, local policy, etc.

Here are some examples of what we mean:

In my work at the U.N. Climate Negotiations I've been working from the frontline of Youth. Most of the "adult" negotiators (from the Northern countries) at these conferences are debating policy that they will frankly not live long enough to suffer the impacts from (and are otherwise insulated from anyway). But young people will still be around to deal with the consequences of bad policy. Young people have been engaging these spaces to say "our future is not negotiable." And after finding that frontline, my work has been to help align youth in working in solidarity with frontline communities. This means communities in the North and the South who are going to be impacted much more than many of the "International Youth Delegates" and who have clarity about the policies they're advocating for. Therefore, it is the responsibility of the youth organizers to act from their position as a moral voice, to support the demands and priorities of the frontline groups who are attending and speaking out at these conferences: the peoples' movements from the Global South, the Indigenous networks and organizations from the North, the Environmental Justice community reps, etc. "Finding the frontline" of youth does not mean pretending that simply being young means you should be in the leadership of the movement, it means you understand your common cause with others and serve the needs of organized frontline groups. This is often a painstaking and difficult process, as many youth voices are not advocating a Climate Justice agenda. Therefore my role is to help amplify and align youth around this agenda.

To be honest, I stumbled into my frontline somewhat unintentionally. I didn't plan it out, but I stuck with it because it made sense given my commitments. My work with the Mobilization for Climate Justice West helped clarify my political analysis, particularly around the importance of community resilience. I grew up in California where punk rock subcultures gave my life foundational meaning. Through punk rock, my relationship to place became politicized, particularly around gentrification and how the subcultural spaces that I find so meaningful in my life can pose serious threats to the viability of communities to be stable through change. Thinking about how I perpetuate gentrification, as well as fight against it, is a central piece of my role in my community. It's one of my frontlines. I've stayed with MCJW because it gives me the opportunity to support similar struggles across the Bay Area (i.e. the Latino/a community fighting gentrification in the Mission District of San Francisco). MCJW has also shown me that gentrification is a housing and human rights crisis that resonates with climate refugees across the world. Living in the Bay Area is a considerable privilege, in that I am able to draw my frontline and align it with community-based organizations already taking up this work with an ecological, systemic lens.

If our task is to *navigate change*, then what we're really doing is assessing a shifting political moment to turn the increasing fractures in business-as-usual *into* interlocking frontlines.

We've begun this booklet by saying that we're writing for people like us, who are trying to align our frontlines and find smarter ways to organize and help build a Climate Justice movement. So far we hope we've clarified who "we" are, and what it means for us to be thinking about our "roles." Next, we need to define what we mean by "Climate Justice."

What an intense writing process this is becoming.

We've been sort of doing it backwards, by writing a booklet's worth of reflections and ideas and then figuring out what connects it all. initially "find your frontline" was a brief paragraph, and now it's clear that it is the central idea that holds everything together!

Thankfully we've had people like Gopal from Movement Generation to talk with us and help clarify and develop the concept.

Totally. Movement Generation coined it, after all. It was really useful to have Gopal clarify that "finding your frontline" is an aspirational idea. Not all of our work fits inside it. It's not even fully laid out here, but we're glad we can open up space for others to build on it and try it on.

We need to jump a little more into the climate justice political framework for it to fully make sense.

CHAPTER TWO CLIMATE JUSTICE

> These are the times to grow our souls. Each of us is called upon to embrace the conviction that, despite the powers and principalities bent on commodifying all our human relationships, we have the power within us to create the world anew.
>
> —Grace Lee Boggs

People are taking action on climate change in all sorts of ways, from federal green jobs bills to international debates around ecological debt to reparations. Since "climate" has become a banner for a wide variety of intersecting issues and problems, "Climate Justice" has emerged as a particular way of understanding problems, solutions, and pathways for change. While Climate Justice is not the only useful framework to engage the ecological crisis, it is especially meaningful when it is used as a *specific* framework, not as a vague marriage of the concepts of "justice" and "climate action."

DEFINING CLIMATE JUSTICE

Climate Justice is not a static concept. It's still evolving as we write this.[27] Climate Justice is a fluid framework developed by social movements around the world, with identifiable roots from the movements whose shoulders we stand on. Instead of writing our own definition of Climate Justice, we'd like to offer a few definitions that others have posed that have contributed to our understanding and the national conversation:

From Environmental Justice and Climate Change Initiative:

> **Roots in Environmental Justice**: "Climate Justice is a vision to dissolve and alleviate the unequal burdens created by climate change. As a form of environmental justice, climate justice is the fair treatment of all people and freedom from discrimination with the creation of policies and projects that address climate change and the systems that create climate change and perpetuate discrimination."

From Demanding Climate Justice section of *Hoodwinked in the Hothouse*, published by Rising Tide North America:[28]

> **Climate Justice as Evaluative Model**: "Climate Justice is a struggle over land, forest, water, culture, food sovereignty, collective and social rights; it is a struggle that considers "justice" at the basis of any solution; a struggle that supports climate solutions found in the practices and knowledge of those already fighting to protect and defend their livelihoods and the environment; a struggle that insists on a genuine systematic transformation in order to

27 For some of the most recent peoples' statements on Climate Justice, see the *Final Declaration of the Social Movements Assembly* at the World Social Forum 2011, February 10, Dakar, Senegal, http://grassrootsclimatesolutions.net/node/955

28 The authors of the piece cited are Movement Generation, Carbon Trade Watch, and Rising Tide.

tackle the real causes of climate change... Climate Justice addresses four key themes: root causes, rights, reparations and participatory democracy."

From Global Justice Ecology Project:

Climate Justice as Global Justice: "The historical responsibility for the vast majority of greenhouse gas emissions lies with the industrialized countries of the Global North. Even though the primary responsibility of the North to reduce emissions has been recognized in the UN Climate Convention, the production and consumption habits of industrialized countries like the United States continue to threaten the survival of humanity and biodiversity globally. It is imperative that the North urgently shifts to a low carbon economy. At the same time, in order to avoid the damaging carbon intensive model of industrialization, countries of the Global South are entitled to resources and technology to make a transition to a low-carbon economy that does not continue to subject them to crushing poverty. Indigenous Peoples, peasant communities, fisherfolk, and especially women in these communities, have been able to live harmoniously and sustainably with the Earth for millennia. They are now not only the most affected by climate change, but also the most affected by its false solutions, such as agrofuels, mega-dams, genetic modification, tree plantations and carbon offset schemes."

From Indigenous Environmental Network:

Four Principles for Climate Justice:[29] "Industrialized society must redefine its relationship with the sacredness of Mother Earth"

1. Leave Fossil Fuels in the Ground
2. Demand Real and Effective Solutions
3. Industrialized – Developed Countries Take Responsibility
4. Living in a Good Way on Mother Earth

We encourage you to read the full descriptions of these principles, and we want to highlight the fourth one, "Living in a Good Way on Mother Earth." Indigenous peoples have offered an analysis around our human disconnection from Nature as a key element of ecological collapse. This approach has been recently advocated inside the United Nations by Bolivia's concept of "buen vivir,"[30] an Indigenous principle of living in dynamic balance with each other and the Earth.[31] As part of the feedback process in writing this booklet, Tom Goldtooth from the Indigenous Environmental Network (IEN) shared with us the Dakota Indigenous concept of "*Mitakuye Owasin*," which means, "We are all related." He elaborates, "It is about humanity restructuring its relationships to the sacredness of Mother Earth, to community and to nature. The industrialized world has removed humanity from nature that requires a need for organizers to be intentional on restorative justice strategies on reframing our relationship to the Circle of Life, Mother Earth and our cosmovision."

---

29 For full descriptions of these four principles, see http://www.ienearth.org/docs/IEN_4_Principles_of_Climate_Justice.pdf

30 http://indymedia.org.au/2010/02/28/the-indigenous-decolonial-concept-of-buen-vivir-in-latin-america

31 This is expressed in the outcome of the Cochabamba Peoples' Agreement and the draft Universal Declaration on the Rights of Mother Earth, adopted by a consensus process of thirty-five thousand people that converged at the World People's Conference on Climate Change and the Rights of Mother Earth, April 22, 2010, Cochabamba, Bolivia.

A few assumptions across all of these definitions are:

1) **Rights-based framework:** using a rights-based framework in organizing means advocating not just for individual liberties, but for *collective rights* of groups such as Indigenous peoples. In the United States in particular, we are accustomed to thinking that individual rights are protected by the government or enshrined in international human rights laws.[32] But Tom Goldtooth explains that this framework has been part of "colonial mindset that has disregarded the self-determination and collective rights of Indigenous peoples, including our right to own and control our lands, territories and resources, our right to free, prior and informed consent, our cultural identity, and other issues." He tells us, "the big push for the last twenty years for Indigenous peoples in the U.S. and globally in our organizing is pushing for our collective rights as peoples (with the 's'), as a fundamental way to have self determination."[33]

Tom gives us this example: The United Nations International Covenant on Civil and Political Rights and *the* International Covenant on Economic, Social and Cultural Rights state that all peoples have the right to "freely determine their political status and freely pursue their economic, social and cultural development." But because the term "peoples" is nonspecific, its application has been in dispute. National governments oppose use of the term "peoples" in regards to Indigenous peoples because they fear its association with the right of secession and independent statehood. To this day, government policies threaten the basic existence of Indigenous peoples. In international discussions, organized Indigenous groups have consistently argued for the development of new international documents addressing the specific needs of the world's Indigenous peoples. This is why IEN and other climate justice activists support the implementation of the UN Declaration on the Rights of Indigenous Peoples (UNDRIP).[34] The Declaration sets the minimum international standards for the protection and promotion of the rights of Indigenous peoples. It is a *declaration* and therefore not *legally binding* unless organizers can support campaigns for the full implementation of UNDRIP within their national governments. Tom warns us that "if the Declaration is not implemented, Indigenous peoples will continue to be oppressed, marginalized and exploited within mitigation and adaptation measures of domestic and international climate change polices."

Gopal Dayeneni from Movement Generation described to us further how a rights-based framework is leveraged in our day-to-day political organizing: "When we do local or regional organizing, we can advance collective rights and unenumerated rights, such as the Rights of Nature or the Rights of Mother Earth, as frameworks for policy and action. Rights do not have to be recognized or ratified by the U.S. government for us to know, intuitively, that they are rights—or to build consciousness, policy, and action in our communities that protect and advance those rights."

2) **Justice as central:** In this context, justice means that communities who have suffered at the expense of our planet-trashing economy do not suffer in order for

---

32 Such as the UN Universal Declaration of Human Rights. December 10, 1948.

33 This was made quite plain to Joshua when speaking to Jonathan Pershing, the lead U.S. negotiator at the UN Climate talks in 2008 in Poland, who said that the U.S. would not sign onto any climate treaty that recognizes Indigenous *peoples* (plural) instead of *people* (singular), because the implications for U.S. domestic policy when addressing Native American land rights and treaties would shift dramatically.

34 After taking more than twenty years to draft and agree, on June 29, 2006, the United Nations Human Rights Council adopted the UNDRIP.

carbon-reductions to occur. Climate Justice is compelling because we do not simply hold up "justice" as a moral obligation, but as a pragmatic pathway forward. Many of the communities who have been left out of "development" have maintained ecological stewardship and lived in balance for centuries. Climate solutions that are not populist and serving the needs of disenfranchised peoples will fail. Conversely, climate solutions that serve the needs of those most affected will benefit everyone.

3) **"Emergency Mode" Trojan Horse:** Climate Justice looks at the scale of the crisis with open eyes. We have no illusions about how urgent and massive our challenges are. Yet, it is also underpinned by the long-view approach to social transformation. The mentality of "urgency and crisis" is often misused as an excuse to ignore justice concerns and make unprincipled compromises that affect the most marginalized communities. In this sense, it was used to justify a Kerry-Boxer climate bill in 2009 that actually made the problem worse with huge giveaways to the oil and gas industry, or the Copenhagen Accord that the United States used to hijack the democratic process in the United Nations Conference of Parties in 2009. In both of these cases, pieces of legislation that actually took us backwards, were pitched as "better than nothing." These arguments fall flat, of course, when you evaluate the destructive impacts of these false solutions. While it's true that "we can't let the perfect be the enemy of the good," Climate Justice clarifies just what "the good" actually means through clear principles that look at real impacts on vulnerable communities on the ground.

"Emergency Mode" organizing informs what is often called "political realism." But fighting for changes that actually *hurt us,* simply because they seem more politically viable, is a losing strategy. Let's look below at a strategic approach to climate justice that unpacks what it *really* means to be "politically realistic."

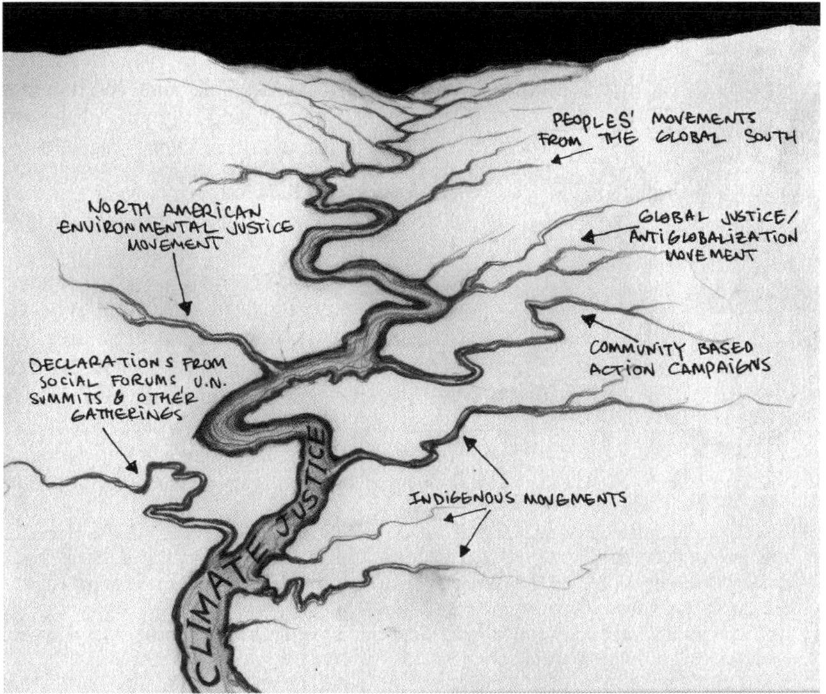

A CLIMATE JUSTICE STRATEGIC FRAMEWORK

A strategy of three circles[35]

Below is one way of thinking about climate justice strategy. The three-circles approach isn't so much "a strategy" as a *framework to develop different complimentary social movement strategies.*[36] One way we like to think about it relates to the relationship between different frontlines acting in different arenas (examples we gave earlier included local policy, corporate accountability, international policy, and local institution building) using different methods: grassroots organizing, corporate and finance campaigning, base building, media spectacle, political advocacy, etc.

Given the current state of ecological crises, this framework can be used to help our organizations, alliances, and movements identify what we believe is materially and culturally necessary (top circle), assess what is currently politically realistic (middle circle), and identify the false solutions that are being put forth by forces with an interest in maintaining the current system (bottom circle). The arrows indicate strategies for making change: winning space to advance our agenda, pushing false solutions off the table, etc.

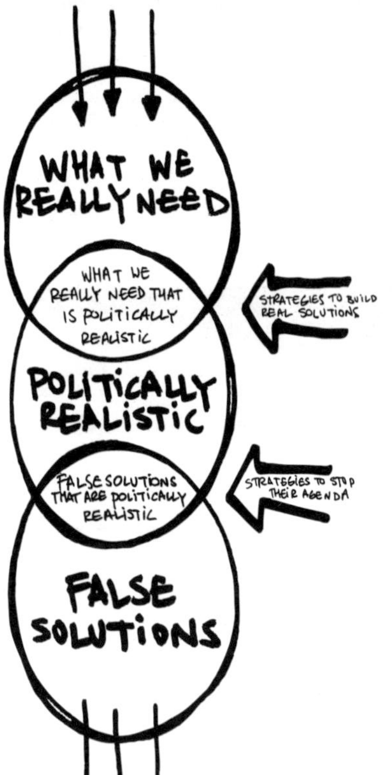

The first circle:

We must identify what is materially and culturally necessary to ensure climate justice. If social movements don't define our visions, we will be stuck in a defensive/reactive posture. Our vision should not be limited by what is politically realistic even if we can't yet win the totality of that vision. Still, some components of what is materially and culturally necessary do overlap with what is politically realistic.

The second circle:

The location of the politically realistic circle is fluid. It is influenced by the balance of forces in whatever arena of struggle you're using the circle to define, which are influenced by social movements, the power of corporations, the current state of the world, etc. It is the role of social movements to move more of what is materially and culturally necessary down into the politically realistic circle.[37] And to push the false solutions out of what is politically realistic.

---

35 This tool was developed by Gopal Dayaneni, Dave Henson, Michelle Mascarenhas-Swan, Jason Negrón-Gonzales, Mateo Nube, and Carla Pérez for the Movement Generation Justice and Ecology Project.

36 See page 48 for more on Social Movement Strategy.

37 "Politically realistic" does not exclusively refer to policy, it can also refer to the viability of *any* objective, like community resource-sharing, winning concessions from corporations, local governance, etc.

The first overlap:

Just because there is overlap between "politically realistic" and "what we really need" doesn't mean that we will get it. And if we only focus on that narrow space, we don't inherently extend our influence to bring in more of our agenda into that space. We have to engage in a variety of spaces to advance that agenda. If we don't say what we really want, we will never get it. So, for example, a total ban on all new fossil fuels exploration and exploitation is not at all politically realistic in 2012, but we don't want to wait until material conditions become so bad that it becomes viable (because at that point, the false solutions will have colonized all the space for what is politically realistic). We have to find ways to advance that agenda now, so that it can become politically realistic. For example, employing strategies that target tar sands as a vulnerable link in the fossil fuels chain, or fighting for protection of pristine ecosystems from new exploration.

The second overlap:

What we really need and false solutions never overlap. What goes in each circle depends on your worldview, what you believe, and the groups you are aligned with.

The circles are not drawn to scale:

The top and bottom circles do not occupy the same amount of space and do not employ the same strategies. Aligning the circles is an important part of the exercise. It allows for a snapshot of the state of play. Then we can talk about what we might move from the top circle into the overlapping space.

Understanding the interlocking frontlines:

Some of our organizations may locate their work (i.e. their organizational mission) in the "what we really need" circle, and not yet in the "politically realistic" overlap. They are working from a place of building out the viability of our solutions, making them compelling and useful, and trying to *push them down* into political realism. Others may be working from the "politically realistic" circle, working on policy, and trying to infuse justice principles in it, thereby *pulling the first circle down.* Others yet might be working on confronting the polluters and false solutions, pushing them out of the national conversation, *or* out of their community. Each of these different locations map to different arenas we work in and approaches we use. Navigating your location is an element of understanding your role as an organizer and *depends on where your frontline is.*

Originally we wanted to go even deeper with this section. There are so many more questions to answer.

The different definitions we offered above are all related, but different. It's a patchwork. We were hoping to engage the ongoing international conversation that is shaping the contours of Climate Justice, including an extensive timeline about the historical roots of it, but realized we were a bit in over our heads with that.

There is no "definitive history of Climate Justice," and we're not the appropriate ones to write it. We're excited to be in conversation with others who are taking this on, though.

Maybe in the next edition of this conversation we can go even deeper. The best way to go deeper is to see how these ideas play out in real life, on the ground.

So lets talk about organizing! That's what aligning your frontline is all about.

CHAPTER THREE ALIGN YOUR FRONTLINE

Organize! Here's the rest of our lives!

—Chumbawamba

ROLES OF AN ORGANIZER

One useful way to think about aligning frontlines to make a distinction between "activism" and "organizing." While different political traditions understand these concepts in different ways, we'd like to offer some that have been useful to us in clarifying our work.

Activism is about using your power and voice to make a change in the world.

Organizing includes that, but in addition is about getting lots of other people to build their power and make a change. Organizing is therefore about *collective action* and working with organized groups. Organizers identify where groups of people are at, meet them there, and work with them in a way that compels them to action.

In our experience, some activist groups who haven't found their frontline all too often inadvertently find themselves unable to communicate their vision, values, and solutions to others outside themselves because they have not prioritized building a base. While we of course want to get our "choir to sing on key," an organizer mindset requires one to reach out to wider and wider circles of people who are not already within the "choir."

If you Give Me a Fish
you have fed me for a day.
If you Teach me To Fish
Then you have fed me until
The river is contaminated or
the shoreline seized
for development.
But if you Teach me
to Organize
then whatever the challenge
I can join together
with my peers
And we will fashion
Our own solution.

Climate Justice organizers working to support someone else's frontline have a number of useful roles we might play, but we must play them with intention and clarity. They include (but are certainly not limited to):

- **Direct support people** work directly for or with community-based organizations in support roles.
- **Movement servicers** offer a particular skill set to different groups, networks, and institutions such as trainers, educators, facilitators, analysts, researchers, etc.
- **Amplifiers** spend time engaging political networks or media to highlight movement work and put pressure on targets. There are lots of traps people fall into playing this role, including taking up political or media space that should be going to frontline communities themselves; that's *not* what we mean here. Instead, this role may include engaging networks, new audiences, or opportunities that are useful but community groups may not have capacity or prioritize engaging.
- **Bridges** work at the intersections of different sectors or movements. This is most clear in our context for those who bridge work with environmental nonprofits and community-based racial or environmental justice groups. Some "bridge" organizers choose coalitions as their political vehicle.
- **Mobilizers** can mobilize large numbers of people in a short-term way in support of community-led fights or concerns, or leverage broader networks to flex their muscles when needed. One example is the new generation of online organizers. Another is community organizers who serve a base in another sector, such as getting labor to turn out in support of an environmental justice fight for a single event.
- **Cultural workers** include artists, musicians, actors, poets, and other creative people who help create cultural shifts necessary for political shifts. This work, too, is often about amplifying stories and requires careful navigation on how it's done accountably.
- **Community organizers** embed themselves in communities directly and even though they are not from the community, they are committed to it for the long haul.

Many of us have overlapping roles within our movement work. Just like using the three circles to help recognize our niche and build our strategy, the arenas we work in determine the most useful roles to play. While there are some challenges that all organizers within the emerging Climate Justice movement have in common, there are also unique predicaments that arise in direct relation to a specific role.

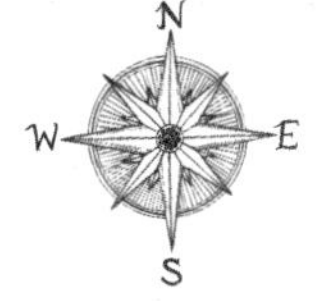

Tuning our Political Compass:

- Many of us play different roles at different times. How do we shift roles gracefully and transparently?
- How do we assess which roles to play and which niche is best for the organizations we build?

MAKING MOVEMENTS, MAKING MEANING

> That is what learning is. You suddenly understand something you've understood all your life, but in a new way.
>
> —Doris Lessing

A critical part of organizing is about creating onramps to collective action. Organizers who have found their frontline have an opportunity to offer these kinds of doorways to millions of others *like us*: ready to take action but in need of some direction to do it in a productive, empowered way.

We do not believe that the central "problem" our generation faces is apathy or ignorance. We think most people in our generation do believe society is broken. They may not agree with us about *why* it's broken, but activists lamenting people's apathy is often a path to our own despair. Yet this idea drives many activists, who then focus on just screaming as loud as possible about the reality of our problems. This approach assumes that because what we say is *true*, it will automatically be *compelling*. As the smartMeme Strategy & Training project[38] says, "Activists assume that because something is true, it will be meaningful to the people we're trying to reach. But In fact, the opposite is often the case: *if something is meaningful, people believe it to be true.*"

Our job, then—regardless of our organizer role—is to help build events, moments, actions, organizations, and groups that *make meaning* in people's lives, and *connects that meaning to taking action.* This requires creating stories that help people understand and name the pain they are experiencing.

Older models of organizing assume that if ignorance is the "main problem" then the task of organizers is just to *fill people with knowledge*. Hence some organizations release endless manifestos, abstract slogans, and try to hammer people with *facts*. But smartMeme offers us the insight that the problem isn't "*what people don't know*" but "*what they do 'know.'*"

Our main barrier isn't that people don't know there is a crisis. It's that they do "know" (believe to be true) that "the climate crisis is a debated theory that may or may not be human-made, that really only affects polar bears, that is mostly rhetoric from Al Gore trying to tax us." People therefore interpret the "facts" they see through that *story*. Instead of convincing them that there are problems, *it is our job to change the story that people use to interpret is the problems going on around them.*

We're not just trying to add new facts; we're trying to change the narratives that people use to understand and *make meaning* from those facts. That's what we mean when we say, "Meet people where they are at."

NATIONAL BALANCE OF FORCES

While the building blocks of our work are local, "meeting people where they're at" also means assessing where our country is at as a whole. With a national or global view, we are able to ask *movement strategy* questions that we don't necessarily see in a local context. Here's an example: many of us lament the massive disinformation climate-denial campaigns mounted by the fearmongers, and the fact that the percentage of the U.S. public who believes climate change "is a major threat and that

38 For more on this see smartMeme's Story-Based Strategy manual *Re:Imagining Change* at http://www.smartmeme.org/change

we can do something about it" has rapidly *decreased* in recent years. While some environmentalists blame the movement's failure to effectively debate climate science in the marketplace of ideas, we think the answer relates to the above idea of *making meaning*. The United States is currently experiencing a highly visible faux-populist right-wing backlash against the Obama administration, fueled by a failing economy. This social movement has a broad *platform*, led by top-of-mind issues like the economy (lowering taxes) and healthcare (privatization). Climate denialism also happens to be a secondary component of that program, and people get pulled into its orbit. The polls are more useful as indicators that a compelling right-wing social movement is offering meaning in some people's lives by providing a *narrative* through which they can interpret the pain they are experiencing.

The Right isn't fighting single-issue D.C. beltway battles, and neither should we. On a national scale, we likely will not see legislative change on climate (as opposed to community-based resiliency projects or local victories) unless it is connected to a larger progressive agenda led by similar issues that feel pertinent to people's lives.

This is another reason why *social justice* isn't just a political conviction we hold, but a pragmatic orientation to be relevant and win.

Tuning Our Political Compass:

- How do we relate our organizer roles to national and international networks[39] of frontline communities and connect their work to broader groups of people?
- How do we engage communities we come from and leverage their power in service of tipping this balance of forces?
- What frameworks do we have to distinguish between our *ideological opposition* (those who are organizing against us because they want a different world than we do—example: the Tea Party) and our *structural opposition* (companies or other institutions working against us because they are pursuing the status quo, regardless of their personal political convictions—example: Coca-Cola). How does this shift the options of places where we can intervene?

COMMUNITY

When we asked Mel Bazil, spokesperson for the Wet'suwet'en First Nation and grassroots advocate for his people, to comment on our section about community, he said, "I see that all peoples face a form of colonization, and that is the struggle, that we all must decolonize. We all must configure our knowledge, our patience, our understanding of each other's struggles, and realize that we are all related and that if some of us are impacted as a human family, than we are all impacted. This means 'not in *anyone's* backyard.' Decolonize the box, don't just work out of that box."

We began this booklet by talking about "finding your frontline" in terms of *community*. The word "community" is often a vague but loaded word that activists love to use at every opportunity. But what does it really mean? Oftentimes, activists imply value judgments in how they speak—that some communities are *real*, and others are not. This isn't helpful to encouraging people to engage those around them, which all of us need to do. Our intention with this section isn't to place a hierarchy of value on

39 Grassroots Global Justice and Right to the City Alliance are examples of national networks that play this role.

different forms of community. Instead, we hope to offer specific and differentiated way to understand the way humans relate to one another so that we can more clearly understand our frontlines.

Communities Are Not Monolithic

Community isn't a romantic, abstract concept that is "good." Instead, communities are specific kinds of social relationships that are the result of social, economic, political, and ecological histories, dynamics, injustices, aspirations, and power relations. There is a lot of internal diversity to any community. Aligning your frontline with others' means thinking critically about the intersections of different *kinds* of communities and how those relationships impact and play out in social movements. Many people who have found their frontline are part of *multiple kinds of communities that overlap.*

Activist Communities

Sometimes people use the term to refer to networks of people working for collective liberation and justice, or a "community" of activists and organizers. It's really important that we see ourselves as part of an activist community, especially if we do so in ways that are inclusive. In this regard, we think of communities as groupings of people with similar commitments and shared work. If done compassionately and consciously, the relationships we build with other activists and organizers helps us bring our politics from the fringe to the center. But this isn't the kind of community we're talking about when we refer to "impacted" or "frontline" communities.

Institutional Communities

Many people build community with the people they share institutional membership with. Perhaps this is a Church community, or a University community, or a workplace. Maybe it's not a physical place, but a *social* institution, like a music-based subculture. We will explore institutional membership (physical and social) in the Social Movement Strategy section.

Place-Based, "Impacted Communities"

Usually, when people refer to "impacted communities" they're talking about a form of community that is *place-based.* That could mean a neighborhood, a section of a city, a rural county, a reservation, or a region, depending on what kinds of commonalities and relationships people have. Most place-based communities involve generational relationships: "Our grandparents knew each other, our parents do too, we know each other, our kids will know each other, and we're all facing similar problems based on where we live." For that reason, communities are a useful way to think of groups of people who face a *common problem* resulting from accumulated social, political, economic, and environmental injustice. They're "impacted communities" because they are impacted by the problem *as a community.*

Other Place-Based Communities

Even though we are all impacted in some way by climate change, we don't necessarily share a common denominator of injustice with our community. Even within similar geographical locations, differences in culture, tradition, and social histories mean that people experience "community" differently. For instance, Hilary grew up in the rural suburbs outside of Sacramento, California. While it was place-based, her upbringing was rooted in the myth of the nuclear and independent family, resulting in an experience that was rather insulated from surrounding ecological and social tensions. This doesn't make her community less or more "legitimate" than others, but it does highlight the nuance we need to clearly name our frontline.

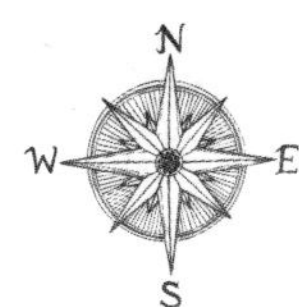

Tuning Our Political Compass:

- What else might community look like that we didn't list here? What's a constituency of people that you feel like you are best suited to organize?
- How does the act of "taking leadership within our own communities" help us "support leadership from most directly impacted communities"? What does that look like on the ground when you don't live near a frontline community (in other words, one that is actively organizing)? What about when you *do* live near one?
- How does our language of serving "frontline communities" and "directly impacted communities" useful or not useful in engaging your community? How can it avoid pigeonholing people, groups, organizations, and sectors of our movement?
- What are equitable (which means fair, not equal) ways to participate in organizing?
- Is "ally" work enough, needed, or always wanted?
- How do people share and encourage the principles of Climate Justice in a community that is not their own while accounting for the privileges that come with that?

Social Networks: The New "Community"?

What about all the people in the United States in the twenty-first century who do not live in place-based communities? Human relationships are changing. It's all too common nowadays that people do not know their neighbors, let alone have peer groups that live in a specific location. How can we understand the new ways people connect to each other? How can we adapt our organizing models to meet our changing relationships?

Much of the new wave of social media attempts to meet people where they're at with this new form of community. Organizations like Moveon.org, Avaaz, and 350.org harness the internet in a way that highlights both its strengths and weaknesses: it has tremendous power for wide *breadth* but usually has trouble with *depth*. Because people have social networks that are not exclusively limited by geography, these organizing models lend themselves to a national or global view. They can therefore assess new ways of taking action and think about the more global conversation in clear ways.

Tuning our Political Compass:

- If many solidarity organizers do not come from place-based communities, how do we evolve our organizing models for the twenty-first century without losing the political insights in older community-organizing frameworks?
- What is the relationship between scale-based internet mobilization tools and deeply rooted work in local communities?
- How do these different constituencies, forms of participation, and mobilization tools differently impact your work, depending on the organizer role you play?

TAKING DIRECTION

Okay, so you've found your frontline and are trying to act in solidarity with a frontline community. But *who* in the community do you "take direction" from?

It's tricky. Communities do not speak from one commanding voice. They're messy and full of people who don't always agree. One way to effectively navigate these tensions is to seek guidance from political leadership. **Political leadership** can take shape in a group of people, or perhaps a few organizations that are trusted by, (and most often from and accountable to) their community.

Political leadership doesn't guarantee one "correct" answer though. For instance, a church can have just as much weight in a community that a social justice organization does, but that doesn't mean they're going to agree on a particular issue. We don't have the answer to such a predicament, but questions worth asking are: "Do we agree with this one position simply because it's an affirmation of what we already thought?" and "Is that relevant to what's actually needed?" All too often activists simply seek out groups within a community that affirm what they previously wanted to do anyway.

The Capitol Climate Action was a mass direct action that happened in Washington DC in 2009. It reached out "beyond the choir" to bring 4,000 people, many of whom had never been to a protest before, to sit-in at the coal-fired plant that powered the U.S. Congress. In an effort to take the wind out of its sails, Democratic Speaker Nancy Pelosi announced that the plant would be converted from coal to natural gas a week prior to the action. This was an immediate predicament for the organizers, because some in the local community that lived around the power plant wanted to declare victory, and initially thought the action could be transformed into a celebration. Yet other frontline communities across the country are suffering because of natural gas exploration and hydraulic fracturing (called "fracking"). Organizers knew that while we were doing our best to serve the priorities of the local community, we were also accountable to a larger politic and movement and had a responsibility to negotiate with the community around clarifying the demands of the symbolic national action. This particular challenge resolved itself through conversation. We put out a press release saying that our protest would continue, and Pelosi's decision validated the efficacy of our methods...but the experience illustrates common complexities of what it means to follow "the" community, particularly for those of us who often help with national and international actions. What would have happened if we couldn't resolve the issue through conversation? National and local priorities can be at odds with one another, and sometimes there aren't easy answers. Sticking to agreed-upon process is sometimes the best we can do to navigate tension.

There isn't necessarily one right answer for challenges like this. In fact, the principles and process of working for Climate Justice have taught us that the best solutions come from collaborative, shared work. Solutions are conditional to a variety of factors in the local environment. The more you get to know and understand that environment, the powers at play, and the people on all sides, the more effective you'll be. In the end, the exercise of taking direction from a community defaults to the degree to which you cultivate *accountable* organizing relationships.

ACCOUNTABILITY

Accountability means that individuals and groups are answerable to their decisions and actions. It also means that even as an individual, you are part of something larger than your own work. We usually only talk about accountability in the negative: when someone is "being unaccountable." That conversation can feel like a field of landmines. This is why we promote *active* accountability, the kind we *want* to take responsibility for. Our good intentions can complicate accountability because, in a field of landmines, it's intimidating to take risks and to innovate. In that way, we no longer need to look at it as black-and-white being "accountable" or "unaccountable" but instead as a path we are all constantly walking as best we can.

Clayton Thomas-Muller from Indigenous Environmental Network (IEN) shares what accountability means for them when aligning with environmental non-governmental organizations: "IEN has...always tried to be very principled about how we work with non-Native organizations. One of the ways that we do this is to insist that these organizations engage with Native communities in ways that are respectful of our unique needs as Native people. We need to be sure that they are not tokenizing our community leaders in campaigns and initiatives that build the profile and power of that particular NGO instead of helping to build the power and profile of the community... We push them to develop mechanisms to make sure that the free and informed consent of Indigenous communities is respected, and to make sure to involve all community stakeholders (I hate this word but will use it for lack of a better one), including our traditional people, our hunters, our women, our youth, and not just the council governments."[40]

Building a practice of active accountability is a core component of *aligning your frontline,* because it can help:

- to build trust between groups that the social, political, and economic powers want to keep divided;
- to be consistent with stated and shared goals;

---

40 From "Just Environmentalism?" http://www.rabble.ca/babble/environmental-justice/just-environmentalism-interview-clayton-thomas-müller

- to ensure that our actions and decisions do not exacerbate existing inequalities and injustices; and
- to build relationships that repair our social relations across difference. So often we've been socialized with destructive behaviors or ideas that undermine our ability to build collaborative social and political power.

Principles of Accountability

Sometimes people who think of themselves as "allies" misunderstand the idea of accountability and think it means just doing what you're told. But accountability is reciprocal and cyclical, it's not something that flows *at* you. Below are four principles necessary to working for accountability.

- **Transparency** means being clear about your politics, organizational structure, goals, desires, and even weaknesses. The point here is to be as open as possible about the perspectives and motivations we bring to begin working from the same understanding.
- **Participation** is about actively and equitably contributing especially in regards to decisions that affect people directly. Most often participation refers to the abilities to contribute to decision-making.
- **Reflection and Deliberation:** the commitment to developing the process of accountability, as it will shift and change in time and with different people. Deliberation means that every part of the accountability process is open for discussion, which account for the practices and mechanisms put in place for accountability, but also the culture and knowledge that surround accountability.
- **Response:** the ability to make amendments, adjustments to issues raised by the Reflection and Deliberation principles.

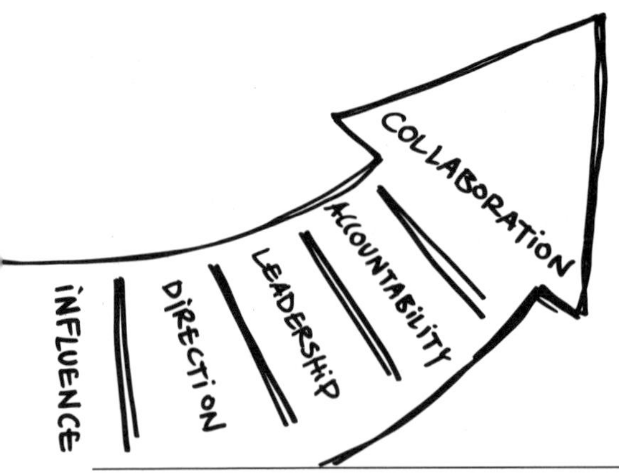

Notice that we've used a circle to represent accountability principles. More than aesthetics, it's meant to represent the fluid relationship between each piece. While we don't think there's a formula to this stuff, below, we're proposing a pathway to frame our approach to accountability:[41]

Influence means finding an organization or a form of political leadership that inspires ideas. This phase doesn't involve a

41 This section is built around a model developed by Rafter Sass of Liberation Ecology Project, with the help of fabulous contributions from Rafter.

direct relationship, or even a specific project. It's more about identifying the sources of our inspiration, our ways of looking at the world, our motivations, and our strategies. If I am starting from a point of not knowing where to look, or who to look to, for leadership and accountability, then mapping out the people, projects, and communities that are influencing me is a way to begin the conversation. This may seem basic, but it is also fundamental. This part of the work deserves discussion, because it's a necessary precondition for everything that comes after—and a perfectly legitimate place to begin! This mapping of influences can help identify missing links, as we ask ourselves: are we being influenced by the people on the frontlines? Are we being influenced by the people we're hoping to be relevant to? If that answer is *not really*, then it's time for some homework. What can we read, what can we listen to, who can we talk to? Starting from wherever we are at, we can begin to get familiar with the perspectives of the people on the frontlines, whether that's in our home community or further afield, and use that influence to guide our next moves.

Developing the ideas generated in Influence, the **Direction** phase involves getting specific about how the perspectives of frontline communities are shaping our work. Like Influence, it doesn't necessarily involve a direct relationship; it can be as simple as responding to a public call for action. By clarifying the relationship between what frontline communities are asking for, and *what we are actually doing in our frontline*, we can identify a mandate for our work. Making the commitment to identifying a mandate, in the sense of democratic authorization for our actions, and letting that mandate shape our decision-making, creates the needed foundation for forging more direct relationships later on.

Jumping off from the mandate developed in the Direction phase, **Leadership** requires *establishing* a relationship, either interpersonally or organizationally. Communication can begin to shape and guide the development of complementary perspectives and strategies. Relationships deepen that encourage participation across groups, producing work directly shaped by the identified group.

Building on the relationship formed in the Leadership phase, **Accountability** involves raising the stakes on that relationship, developing and deepening trust. Becoming answerable for our actions and getting feedback are the critical components of this phase. And through the trust we build, accountability means that the groups are beginning to share priorities, in work practices and political vision, as well as interpersonal values.

- **Collaboration** means that the partnership has developed to actively and cooperatively create original work.

  Points to take home:

- **Standard of Accountability:** Notice that accountability isn't the end goal; rather, the end goal is collaboration. This is because accountability should be a standard rather than distant objective. In this way, the pathway proposed displays the arc of collaboration among different frontlines.
- **Practice:** Take some time to honestly consider where your project or organization is located on the pathway. What necessary steps can you identify to help move toward collaboration?
- **Cumulative:** Hopefully it's obvious that working for accountability, toward collaboration focuses heavily on relationship-building. This also means that working for accountability is a cumulative process that can't be done overnight, although there might be substantial, quick actions that can be taken now.

- **Collective:** Do you know *why* you'd want to build relationships across different sectors of the movement? How does it move the work forward as a whole? What is the strategy in aligning frontlines? This question is useful to break away from the "everything is fixed by having relationships" stance.

Obstacles We Face

Taking these ideas seriously means confronting a lot of assumptions, insecurities, and sometimes lies we tell ourselves. We also believe that each phase is going to be harder than the one before. In fact, truly collaborative work is the work of creating new sets of rules, new kinds of social relations—basically creating the world we know is possible. In our booklet feedback process, Rafter Sass from Liberation Ecology Project shared this:

> Just because accountability is our watchword, and collaboration is our ultimate goal, doesn't mean that influence, direction, and leadership are second rate. Each phase of the relationship includes, builds on, and deepens the foundation created in the phase before, while it creates the conditions for what is to come. We can start wherever we're at, and feel good about our work, and at the same time keep the goals of accountability and collaboration firmly in our sights. And we have to, because at this moment in history, influence, direction, and even leadership just aren't enough. The inequality we face is so profound, and the wounds in the social fabric are so grave, that nothing less than a new way of making change is required of us. We all have to work together across the boundaries thrown up between our communities, between power-holders and those who have historically been disempowered. And before we can truly collaborate across the boundary, we've got to put in some work. Those who have been granted power by this system must become accountable to those who have not.

That's not to say that accountability and collaboration aren't possible right now. Quite the opposite! We're building pockets of this work that exist in tension with the status quo, the bigger world. Navigating that tension is part and parcel to finding and aligning your frontline. What's more, we have to find ways to translate these skills and ideas out to more people, empowering them to find what's relevant, beyond our political circles.

With that said, some of the most common obstacles within this work, that we've faced, come from being uncertain and not having space to talk about this stuff candidly. This is a goal of ours: to start talking more intentionally about what accountability *is* rather than *isn't*.

Inactive and Active Accountability

We began this section by writing about the need for "active" accountability that we *want* to take responsibility for. Sometimes working for accountability can be a scary exercise in simply avoiding criticism, which inhibits risk-taking and stunts movement growth. In order to foster a culture of active accountability, we can build practices that avoid some of the pitfalls groups commonly trip over:

Moving From Inactive to Active Accountability			
INACTIVE	**DESCRIPTION**	**BRIDGE**	**ACTIVE**
"Tell us what to do"	A meek reaction to "taking leadership from frontline communities." Possible sources: feeling defeated or uncertain.	**Take yourself seriously.** Likely you'll feel conflicted as to what your role or contribution is to this work. This means having confidence in your skills while maintaining lightness to your seriousness.	**"What is our strategic political moment?"**
"Just trying not to step on any toes"	As long as my work doesn't make anyone mad, I can do what I want.	**Skill share!** If you're working more or less independently from collaborative projects, try exploring how your skills, if shared, could amplify collective struggles.	**"In our political landscape, this is what we have to leverage"**
The dreaded check-box	This happens when a fair amount of organizing takes place without all necessary players, and people then "check in" too late. Often, prioritizing logistics over accountability gets us here.	**Pump the brakes.** Understand that not everyone is on the same timeline and work collaboratively to meet in the middle. It's better to have a project on hold than a project that stunts existing work.	**"Now that we're all here, let's map out our options!"**

EXPERIMENTS IN ACCOUNTABILITY

So what would it look like if we took these ideas and practices seriously? In our experience, that requires a considerable amount of experimentation, not just at the individual level, but at the organizational level, too. We'd like to offer Hilary's experience with Mobilization for Climate Justice West (MCJW), written in the first person, as a case study that looks at our organization's structure as a model for experimentation. MCJW is a regional *alliance* that involves a wide variety of groups working for climate justice in the Bay Area. We'd like to clarify that MCJW has since adapted a new model, but still want to use this example because it conveys a deep commitment to two contentious but critical concepts: *accountability* and *capacity*.

After one year of intensive action organizing (seven large-scale actions in five months), MCJW decided to reflect on the recent work to figure out our next steps. In February 2010, MCJW held its first organizational retreat. A lot of time and energy was put into careful reflection on past events, future goals, and practical next steps. It was here that we decided to be more explicit in our priorities as an alliance, namely to further internalize our commitments to the practice of Climate Justice principles in our internal dynamics and decision-making.

Moreover, we needed a way to make sure our goals, principles, and alignment were at the forefront of the work. The biggest tension we faced was between accountability and capacity. We acknowledged that people participated in different ways, and that this fact required a heightened, constant awareness and responsibility in our practices and organizing culture. For us, it seemed that people with the most time and energy to contribute weren't connected to an organized community or a base of people. In fact, most of us were young, white college students with a lot of good intentions. Moreover, the people who worked for an organization that served a frontline community often had the least amount of time, as they are devoted to their own campaigns.[42] There are other layers, too, that complicate this situation, like: should people that get paid to do organizing work take up more tasks? And how much should our organizational structure try to shape what accountability looks like?

And so, we created a unique organizational model, pulling from various other examples, to really try and meet our needs and goals.

The first mechanism for accountability was requiring membership be based on *organization*. Individuals were able to participate, but had no decision-making power in terms of setting big-picture organizational strategy. The intention here was to encourage individuals to organize themselves with groups that they are accountable to. Membership operates then, under two assumptions: 1) only organizations are members—there are no individuals with MCJW membership; and 2) the organization has agreed to our stated goals, principles, and alignment.

The second mechanism for accountability was the Coordinating Council, which included:

- At least 50% representation from *community-based* member organizations.[43]
- Ensuring accountability and visibility to local campaigns by strategic planning process and general administrative needs.
- Proposals for campaigns or events either beginning with the Coordinating Council, or have to go through the Coordinating Council

Below was the general pathway of proposals. Working groups came up with an idea which was vetted by the Coordinating Council and, if approved, went onto the organizational member spoke council for adoption.

During the February retreat, we agreed that we'd come together in six months to discuss the viability of this structure. We learned some really important lessons. This level of experimentation distilled our strengths, weakness, and complex situations that we weren't previously aware of. I think this level of experimentation was possible

42 For example, grassroots organizing against Chevron Corporation's Richmond Refinery.

43 Definition of Community-based Organizations (CBO), written by MCJW: Funded organizations, volunteer-run groups, or collectives and their networks: A) who engage in direct grassroots organizing in a frontline community and takes direction from and is accountable to that base of members; B) whose primary mandate is to work with, provide services for, and empower local communities; C) that spend more than 80% of their time working on local community-based initiatives.

for a variety of reasons (culture of political expectation in the Bay Area, long-term relationships and work already built around campaigns against Chevron, concentration of skilled veteran organizers), but overall it illuminated the tensions that exist within the larger climate and environmental movement. Ideally, the process of this experiment is useful to reflect upon your own project and organization. How would taking issues of accountability and capacity seriously impact your project or organization?

Big-Picture Assessment of MCJW's Structure Experiment

POSITIVES	CHALLENGES	LESSONS
A serious **experiment** in trying to equalize the inequalities that exist in this work and in our society.	**Mechanized Fear**. Admittedly our structure was created from a place of fear, trying to account for all possible ways people can be unaccountable.	Building a **culture for accountability** is a must. You can't necessarily structure accountability because it is ultimately about trust and relationships. Yet, there are practices that organizations can undertake to foster accountability.
A **useful model** to build from. The lessons learned from this experiment directly informed our amendments to MCJW's organizational structure.	One lingering tension was the role of "individuals." Lack of **capacity** was a residual challenge that cannot solely be addressed by an organization's structure.	Processes to work for accountability have to be **accessible**, but that isn't a given. Accountability processes require **commitment** and direct **communication.**
This process **strengthened relationships** among those involved, especially cultivating a culture of responsibility and trust.	**Too many meetings!** Adding layers of accountability mechanism, in this case, required more participation for those with limited capacity, and split the capacity even among those with more time, somewhat exacerbating an inequality we were trying to alleviate.	A willingness and ability to **take such a risk**, organizationally and interpersonally. It's essential to continually ask "What is most relevant?"

RELATIONSHIP-BASED ORGANIZING

Regardless of approach, framework, or role, all organizing is about building relationships. In our experience, the above questions become easier to navigate through the process of creating strong and meaningful relationships with fellow organizers and community members. Yet, this is also one of those things that need to be unpacked, because "building relationships" doesn't just *happen*. It takes a lot of work! Below are a few essential tenets to building personal relationships in political projects, particularly across frontlines.

Respect Then Trust

Since the movement for Climate Justice isn't trying to be insular, we have to finds ways to build trust with people who might not always be easy to build with. We can't simply hang out with our friends if we're trying to build a movement. So, let's break it down a bit:

- **Respect:** The positive feelings and actions informed by positive esteem for someone.
- **Trust:** Having faith that someone will do what they say they are going to do; relying on another person, or believing in them.

Having respect and trust for people are challenging things, in general, let alone for strangers or people you disagree with. Human beings are often exasperating creatures. Sometimes it takes a lot to even be convinced that you should *try* to build those things with someone, let alone keep at it.

One of the simplest ways to begin building trust is demonstrating respect. One way to do this is to discover the art of *active listening*, that is, stop talking and start hearing people's stories and ideas. A useful way to check if you're practicing active listening is asking yourself the question: *Do I formulate my response while someone is still talking*? You can also "WAIT" before opening your mouth (ask Why Am I Talking?).

Another way to build respect that moves toward trust is speaking truth to power dynamics, both real and perceived. This means getting real with the fact that we are not all treated equally, and pretty often people don't feel like "we're in this together."

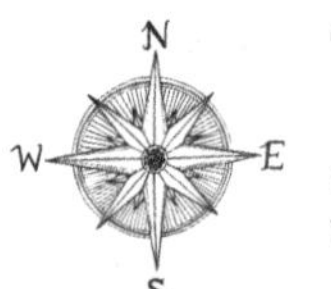

Tuning Our Political Compass:

- How do we wield privilege without abusing it, particularly in coalition work? How can we see that line before we've crossed it? What would that look like?
- How can I become comfortable constructively disrupting the status quo, especially in a room of "like-minded" people?
- What do I need from myself and others in order to take risks in speaking truth to power dynamics?

Taking the risk to speak truth to power dynamics is essential but is a double-edged sword. While there is considerable, unique knowledge gained from lived oppression, this can sometimes lead to activists (particularly solidarity organizers) competitively accruing "points" or badges of legitimacy by pointing out oppressive internal dynamics. This is not useful or healthy for group dynamics and can create a similar "field of landmines" dynamic illustrated in the previous "accountability" section. Feminist thinker bell hooks also reminds us that "oppression is the absence of choices."[44] Our job in speaking truth to power dynamics is to *increase* the choices we have as individuals and as a group together.

In order to move from respect to trust with someone, it has to be reciprocal. *It doesn't work if only one person or group respects the other*. But what if you *don't like* somebody that you need to build with? Our answer: it's important to remember that organizing isn't about *you*. Frankly, there are plenty of people who do solid work and deserve our respect, even if we find them personally distasteful or obnoxious. Respecting people is a *choice*. It's important to check ourselves on why we make the choices we make.

Plus, the fact that organizing isn't about you:

---

44 bell hooks, *Feminist Theory: From Margin to Center* (Boston: South End Press, 1984).

- 1) Is a relief—whew!
- 2) Puts an exciting spin on things: If I'm capable of offering my respect, when it's hard, so that I am open to building trust with people, that means we're building a different kind of relationship—an unconventional, even intentional relationship. On a regular basis, we hear people refer to each other as brother and sister within our organizing. These terms of endearment establish something really powerful: it means we can grow together, like a family. We don't always have to agree, in fact, we grow the most when we have the space to challenge each other and it's because of these relationships that we can innovate our work.

Consistency

Organizing means taking the time to understand the landscape to which we live and do our work: the history, the politics, the social dynamics, and the ecological terrain—all of it. Building real knowledge and experience around that stuff takes a considerable amount of time, usually measured in years not months. Intentional relationship with a place means investing even when things aren't buzzing with excitement.

Mistakes and Uncertainty

Most of us aren't very graceful in our feelings and behaviors that involve mistakes and uncertainty. In fact, we're often taught that if you can fake it, you can make it. That makes it difficult to learn the skills needed to cultivate humility. Sometimes aligning our frontlines across difference is intimidating, and we want to appear confident and competent. Or other times we're pushing hard against forces that are destroying communities, that we can become self-righteous as if we had *the* correct answer and everyone else is wrong. Finger-pointing and criticizing other movement actors is often much easier than looking in the mirror.

Getting to know ourselves in the moments that push us to an edge is really important in this work. It's also connected to the whole "consistency" aspect—you stick it out—which is really hard if and when you mess up. We develop a lot of defense mechanisms (becoming aggressive, evasive, submissive, or suddenly acquire a case of amnesia) that stunt this process which is ultimately about trying to understand what's going on under the surface of raw emotion, like:

Past trauma: The trauma experienced in our lives manifests in sometimes unexpected ways. Not to get super psychoanalytical, but looking to past events and interactions is helpful in understanding what comes up. This is the same for everyone, in that, it is important to remember *that in difficult moments you're often interacting with someone else's trauma and coping mechanisms.*

Social Histories: There is also trauma in our personal social and cultural histories enacted by legacies of domination. We often bring this stuff with us (both as survivors and perpetrators of domination) when we walk through that meeting door. It often manifests when collaborating across frontlines. Our sensitivity to these legacies is especially heightened when we feel like it is being compromised or devalued. Looking to past events and interactions are helpful in understanding what comes up.

Feedback

Making mistakes and having uncertainty is best remedied with feedback. But not all feedback is the hand-holding kind. Some kinds that haven't been useful to us include:

- *On blast* – Have you ever been, or seen someone "called out" in a meeting or a public setting? Situations like this can occasionally be necessary, but they also make it really difficult to receive an important message and often create a destructive internal culture.
- *P.A.A.S.* (Passive Aggressive Aggravating Statements) – Passive aggressive comments are aggravating because they can be so ambiguous in intent, delivery, implication, and assessing how to respond.
- *Shit Sandwich?* – This is often taught as an organizing tool. It happens when the more difficult feedback is sandwiched between two forced compliments. While it offers a thoughtful padding, it actually complicates our relationship to receiving positive encouragement. One similar alternative is the 80/20 principle: that you should give 80% positive feedback and 20% criticism. We don't have a particular opinion on the ratio of feedback, and we do think *affirming feedback is important* to hear alongside criticism, but we've seen dishonest or forced compliments make things worse.

These forms of feedback are full of assumptions and awkwardness, which is all the more reason to take initiative in our mistakes and uncertainty by seeking out honest and constructive feedback.

- *Ask for it* – Asking for feedback often may be a simple idea, but it's also a tall order. Giving feedback can be just as hard as receiving it. We are not only tasked with taking the initiative to ask for it but should do it in a way that is accessible for people to respond thoughtfully and honestly.
- *Give it* – Directly, clearly, honestly. Feedback is reciprocal. The more you give someone feedback, the more you strengthen relationships and make people comfortable giving you feedback in return.

It's important to remember that all forms of feedback hold granules of useful information that sometimes you have to dig for. The task is developing skills necessary to distill the feedback, which usually depends on how it was transmitted. A useful analogy that I like is the owl pellet: digest what is useful and spit up the rest.

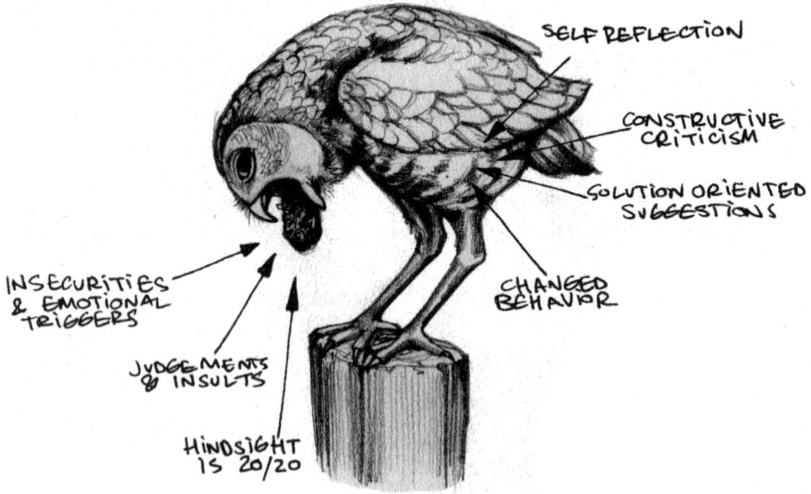

Intentional relationships

One thing that makes it easier to get comfortable with our mistakes and uncertainty is building intentional relationships with people. Intentional relationships are integral to personal and political growth because it means that you have a support system in place to weather difficult situations.

Impact vs. Intention

How is it that good intentions can easily go bad? "Bad" happens when our intentions take center stage while we distance ourselves from the impacts. In other words, using our intentions to skirt responsibility. Having communicative, supportive, intentional relationships with people helps us move away from taking action based on intention and moves us toward taking action based on shared knowledge and collective goals.

Reconstructing Our Social Relations

We all have that friend who is really smart, an expert in their field, a valuable asset to this political work, but is rather awful at interacting with others. This isn't a jab at socially awkward people; it's a shot at those who put their politics above their social relationships. Intentional relationships are the opposite of that. It means that we put our relationships at the center of the work, and *the integrity of the relationship between people is as much of a goal as the political objective.*

Mentorship and Peership

> I believe in the right of people to expect those who are older, those who claim to have had more experience, to help them grow.
>
> —Ella Baker

Explicit mentorship often arises when I seek advice and support as a younger, less experienced organizer. For example, when MCJW was planning our February 2010 organizational retreat, it was clear that we needed to create a proposal for a new organizational structure. Carla Pérez, from Movement Generation, Sharon Lungo from the Ruckus Society, Ananda Lee Tan from the Global Alliance for Incinerator Alternatives, and I held a meeting about the retreat; it was also decided that I would begin drafting the structure proposal. At this point I had neither experience creating organizational structures, nor the awareness necessary to incorporate a multitude of factors, like the terms and agreement members employ, let alone identifying what concrete contributions "members" should make. In struggling with this task, I asked Carla Pérez and Sharon Lungo for help. They not only encouraged me in this process, but also offered tangible resources, like the internal frameworks of organizations they previously created, and many one-on-one brainstorming sessions where we were able to discuss and create the organization's structure proposal.

Through our feedback process, Carla Pérez from Movement Generation shared what intentional mentorship looks like:

> It is important that we cultivate and nurture learning relationships where we not only share information with each other about the issues we are taking on, but that we also build relationships that lead to developing new leaders. This involves critical subtle skills such as how to respond to difficult group dynamics or problematic participants, how to frame things in a positive way towards resolution, and how to ask questions that don't make others feel defensive but still hold people accountable. And I've seen this in MCJW with some of the challenging "structure" conversations regarding certain individuals' participation. That can be something that those of us who are older in the movement *explicitly talk about as part of what's important in organizing,* because it is. We can be conscious of it and remind each other about the various ways to handle disagreements.

Another intentional relationship experienced in MCJW is *peership*, or affinity based on similar organizing experience and sometimes age range.

Cathy Kunkel from Rising Tide Bay Area explains this idea further: "It has been really helpful to navigate challenging situations with peers who are, like me, still finding their place in the movement and figuring out what skills and talents they can contribute."

Cathy raises the importance of having *different kinds* of intentional relationships. This is another crucial point in the processes of repairing social relations. Activists and organizers experience a plurality of relationships in political organizing work, not all of which are sustaining "fuel." In fact, many are not. Because "we experience our politics through our social experience," as Carla Pérez says, we need to have relationships that provide a sense of trust, guidance, and intentional commonality.

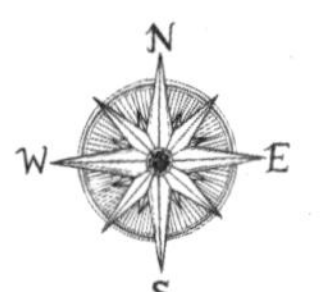

Tuning Our Political Compass:

- What are our motives for relationship-building? Is it something we just have to do to do the work? Are we cultivating meaningful interactions that support our everyday lives?
- How do the relationships formed in our political lives intersect with our social, personal lives?
- In what ways do these relationships frame what is needed to build more resilient communities to weather the crises we face?

CARE AND TRANSFORMATIVE PRACTICES

> We go to demonstrations, we build events, and this becomes the peak of our struggle. The analysis of how we reproduce these movements, how we reproduce ourselves is not at the center of movement organizing. It has to be.
>
> —Silvia Federici, feminist scholar and activist[45]

45 Silvia Federici, "Precarious Labor: A Feminist Viewpoint," http://www.scribd.com/full/3108724?access_key=key-27agtedsy1ivn8j2ycc1

Our vision of climate justice is one of expanding and interlocking frontlines. But this framework only makes sense when it's based in the nurturing and support that sustains movements themselves. This "invisible" work of taking care of ourselves and each other is often not recognized or valued. Yet its key to "reproducing our movements." Interlocking frontlines means that we need to be building affinity with others, and that means supporting each other in a meaningful ways, especially when it's hard. Therefore, we need to extend our care practices beyond our personal relationships or even our immediate political networks, while understanding that if we cannot take care of each other, we cannot do this work. Care is the basis with which we can negotiate each other's frontlines.

Revaluing Care and Daily Work

In grassroots organizing, it's easy to mark our miles by large-scale events and message-boosting media opportunities. Take this story for instance: In 2009, Mobilization for Climate Justice West took on a project of generating "street heat" in the months leading up to the Copenhagen UN Climate Negotiations. MCJW organized seven large-scale actions and events, including a three-day climate convergence in Richmond, California, where people from all over the West Coast came together to create local solutions to the climate crisis.

Thrilling, perhaps, but measuring success though these "crescendos" doesn't paint a complete picture. The above snapshot doesn't reveal much about the on-the-ground organizing—meeting coordination, material runs, art parties, outreach coordination, grant writing, etc.—that made those events possible. What's more, it doesn't even come close to understanding or cultivating the skills necessary to maintain, build, and repair the relationships that even allowed the unseen organizing to happen.

December 7, 2009. It was the night before our sixth action in the MCJW "Road to Copenhagen" project. I could barely keep my eyes open as we once more went through the plan to deploy at the front gates of Chevron's headquarters, lockboxes in hand. I remember the excitement stirring between the people around me. But we were absolutely exhausted. I was running on fumes and knotted nerves. We were burnt out. It wasn't till a few months after this moment, after our sprint came to a close, that I was able to reflect with my peers and on our year of intensive actions and how it affected us. That moment, the night before the action, is important for me to remember because it was pivotal in changing how I understand what organizing means, what behavior and ideas I prioritize, and how I engage in political work.

All too often, I checked my "private life" at the door of my "public life." Take for instance, some of my most basic everyday needs: food, health, and sleep. Regularly, I found myself so involved in "the work" that I would be up till four a.m., having forgotten to eat. One time, I even drafted a press release while sneezing and coughing up florescent phlegm from a horrendous flu. I put the political work before myself, and my needs. What's more, I didn't think twice about doing otherwise.

We all know that "self-care" is crucial to stay in this work, but sometimes it's easy to see care as just another item on the "to-do" list, which can easily be deprioritized. Jessica Tovar, an Oakland community organizer from Communities for a Better Environment, talks about the tension between care and political work: "I feel like [care] is something I need to sit and think about more... because every time we do take time off, it's out of 'my head is about to explode!' and that's a really bad habit that we've gotten into." Jessica raises a critical and common point in political spaces: taking care of ourselves usually means taking a step back from our work.

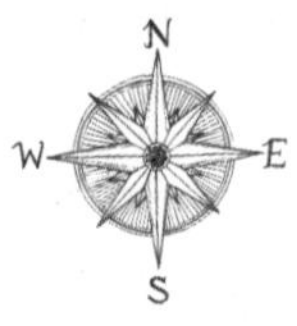

Tuning Our Political Compass

- Why is the practice of removing ourselves from the work often seen as care?
- What about our organizations encourages this? What in particular about how we resource our work, our organization's structures, internal cultures contribute to it?
- How do burnout cycles affect our ability to lifelong effective organizers? How do they affect our very ability to have a long-view of the work?

Our self-care practices can also shape our relationships inside and outside the work. Despite the burnout intensity of organizing that happened during the "Road to Copenhagen" campaign, the relationships we formed felt like a family. In our movements, we often navigate taking care of each other as a chosen family. Here, we learn conflict resolution, intentional communication, setting boundaries and expectations, and a practice of care. These skills strengthen *all* of our relationships, not just how we work on our frontlines.

Relational Practices

In our "biological families," our day-to-day experiences teach us how to be in the world. The same is true for our "chosen families." The "Road to Copenhagen" project was a lesson in how to burn out. With an urgent focus on short-term goals, we had an "all or nothing" approach to organizing.

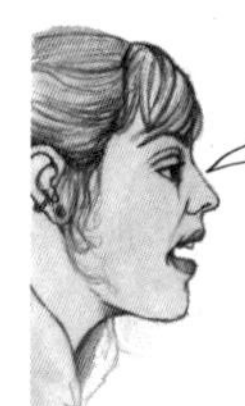

On one hand, large numbers of people were coming together to offer their time and energy, many young white college students who hadn't done this before, myself included. The exposure served as a fantastic crash course in Bay Area politics and decentralized organizing. On the other, this approach limited participation from people of color–led, community-based organizations because their limited resources are often tied up in intensive organizing like base-building, service work, and civic engagement. It wasn't in the interest of these organizations to run full throttle on these events, especially if the events weren't bringing capacity to their work. These gaps became really problematic for a coalition whose goal was to work accountably and collaboratively, connecting international climate struggles with local ones, yet our practice invited burnout. Bummer.

This tension became the focus of an MCJW retreat at the end of that year. Yet, one thing many of us didn't realize till much later was how much that crisis-mode* style of organizing encourages and thrives off individual personalities. For example, there were nine of us at the "core" of MCJW, each with a particular skill set (outreach, media, art, etc.) that created the scaffolding of each event. I don't think this way of organizing people is inherently bad, but without long-term strategy, it tends to be insular.

*Crisis-mode organizing is a concept borrowed from Chris Dixon's "Against and Beyond: Radical Organizers Building Another Politics in the U.S. and Canada."

Unfortunately, it seems we're much less able to care for each other in the face of looming obligation. In this example, MCJW's obligations were determined by a regimented series of actions that generated an emergency-based approach to organizing. Everything was a sprint. Everything was an emergency. As a coalition, we burned out our member organizations. Our private lives suffered because of the urgency of our work lives, and when we couldn't meet our everyday needs, our political lives suffered. After that intensive year, MCJW had to work hard to regain respect, trust, and reclarify priorities with the community-based organizations with which we first worked. That's why a baseline of care is so key when aligning frontlines. We learned that our strategic planning and political priorities shaped whether we were unintentionally using an "emergency-based" approach.

Obviously, legitimizing some work as "real" and keeping other work unacknowledged isn't useful to movement building. Feminist scholar Maria Mies discusses this idea in the context of our economy and its effects in gendering and devaluing labor. Mies says the labor often valued as "smart" or "developed" always requires the labor of marginalized peoples to be hidden, devalued, or stolen. Here, she's speaking directly to the "progress" of the "developed" Global North at the cost (resources, labor, culture, spirituality) of those in the "underdeveloped" Global South.[46] Let us consider some of the most common and marginalized "hidden" labor in the United States: domestic work. Of the 1.8 million domestic workers in the U.S., 95% are female, foreign born, or persons of color. Their labor allows for their employers to engage in "public" work.[47]

Political organizing isn't immune to replicating this tendency that values particular kinds of labor and activities over others.

The ability to "reproduce our movements" requires that we revalue hidden labor and develop useful, strategic ways to take care of ourselves and each other. The Rock Dove Collective, a community health exchange in New York City, explained to us: "We connect those practitioners, who we call 'providers,' with service seekers, ideally based on mutual aid... That means people clean peoples' houses and get acupuncture in return. Or they pay a sliding scale that really does reflect their ability to pay not only including their income but also including how many kids they're supporting, whether they're supporting their elders, that sort of thing."[48] Experimentation in this way is useful because it strengthens and uplifts the ways we reproduce ourselves. Indeed, doing justice to our reproduction requires that we build our movements around it.

Reproducing our movements also means taking up roles that aim to move our communities through these large-scale crises happening all around us. Michelle Mascarenhas-Swan, strategy initiatives director from Movement Generation, says that "The transition is going to take all kinds of people participating and playing key roles—counselors, healers, coaches, organizers, facilitators, mediators—we

---

46 Maria Mies, *Patriarchy and Accumulation on a World Scale: Women in the International Division of Labor* (London: Zed Books, 1998).

47 http://www.excludedworkersreport.org/report

48 Interview with Rock Dove Collective conducted by Ben Holtzman and Kevin Van Meter in Brooklyn, New York, August 15, 2010.

need to learn how to cultivate, support and recognize these roles as key transitionary, green roles." Here, Michelle raises the idea of transformative work, which is the work that helps us recover from dominant culture, learn new ways of relating to other people, and construct our agency to organize with others across frontlines.

Transformative Political Work

An exceptional example of transformative work is captured by Mujeres Unidas y Activas (MUA).[49] MUA engages in healing work that helps members shift from "victim" to "survivor" to "agent for change." One way MUA does this is peer counseling workshops. Members receive training for crisis intervention and reconciliation, which they bring into their homes and community, creating social power. Other MUA offerings include: mutual support meetings, informational workshops, counseling, and job placement referrals.

MUA is based in San Francisco. They are a grassroots base-building organization made up of Latina immigrant women. They've adapted their organizational infrastructure to meet their needs and goals, which are to empower their base through personal transformation and building community power for social and economic justice.

Care and Movement Building

Taking cues from organizations like MUA, those of us who are finding our frontline in the movement for Climate Justice must also draw that line around the work of caring for others.

The intersections of our politics and personal lives are intimate, and if we're not intentional, we can reproduce the dominant patterns of devaluing care and making it invisible. Interlocking frontlines means taking care of each other especially through hard times. If we're building across differences and frontlines, we must also extend these practices of care outside our political networks.

Climate Justice is about cultivating alternative economies, finding ways to organize our communities equitably and democratically, and building people-power. Care is the thread that runs through this work with skills like attentiveness, responsibility, responsiveness, and ability to meet people's needs over the long haul, through transition, and toward more resilient communities.[50]

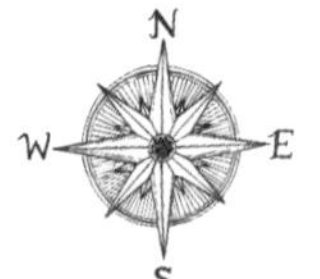

Tuning Our Political Compass

- How do the "care roles" outlined in this section relate to the "organizer roles" in the beginning of this chapter? How do you move between them? Center them in political work?
- How does self-care, when taken because it's an emergency, impact you, the people around you, and your political work?
- Does your political project prioritize personal transformation, through organizational infrastructure, or otherwise? If not, what might that look like?

---

49 http://www.mujeresunidas.net/

50 Characteristics of care found in Joan Tronto, *Moral Boundaries: A Political Argument for an Ethic of Care* (London: Routledge, 1993).

- What does the practice of collective, intentional care look like to you? How is that different than extraordinary care, like spas and vacations?
- How can you strengthen and amplify care practices in your life, the life of your loved ones, and your community? How might this be a collective effort?

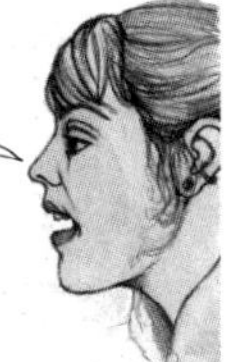

CHAPTER FOUR
TAKE ACTION

Strategy asks the question, "What can we do today, so that tomorrow we can do what we are unable to do today?"

—Paulo Freire

MY QUESTION IS: ARE WE REALLY MAKING AN IMPACT?

SOCIAL MOVEMENT STRATEGY

What distinguishes Climate Justice organizing from the broader climate movement is its *systemic analysis* that we visited in earlier chapters. Climate Justice has an analysis of the "invisible" social, economic, and political institutions that bind us all together and shape our society. It is this systemic analysis that tells us that individual consumer-based action is good but not enough, and that we therefore need to build movements and organize collectively. Communities that have found their frontline often go through a process of being able to name these systems that structure our lives.

But while many activists can *analyze our problems* systemically, we often do not have the tools to map out our *movement building process* systemically.

Social movement strategy moves beyond the narrow lens of a campaign and maps out human relationships to shift the balance of forces in a whole society.[51] All people are connected through groups and associations. Some of these are physical

51 For a refresher on campaign strategy, see page 50

institutions (unions, churches, schools, etc.), sometimes they are invisible ones (like youth subcultures such as hip hop or punk rock).

Successful movement building often hinges on being able to see a society in terms of specific overlapping networks of relations between people. That's why we did the work the previous chapter of defining a few different forms of community. When a constituency is engaged *as a constituency*, social movements have an opportunity to shape people's worldviews. Some of the movement leaders we know best were successful because of their ability to mobilize different constituencies. For example, Martin Luther King Jr. was able to *build a base* through engaging the Southern Baptist Church. He engaged that institution in such a way that if you were a Southern black Christian at that time, you were likely an agent of change. For many, it was *part of their identity*.

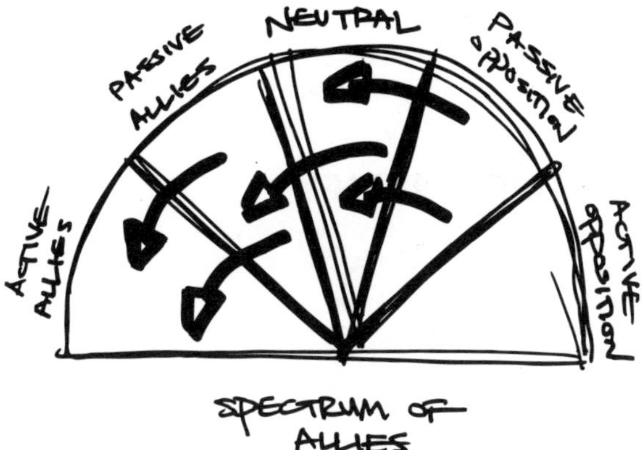

We are able to understand the specifics of interrelated networks that make up different social movement *sectors* once we have a movement strategy lens. We need this lens to see interlocking frontlines clearly and make a strategic assessment about how to move as a movement.

One tool to try to look through that lens is called a Spectrum of Allies.[52] It can be used either to map out a campaign or to strategize a whole social movement.

Here's how it works: in each wedge you can place different people, groups, or institutions. On the left side are your *active allies*: people who agree with you and are fighting alongside you. Then your *passive allies*: folks who agree with you but aren't doing anything about it. Next are *neutrals*: who are like fence sitters. Then *passive opposition*: people who disagree with you but aren't trying to stop you. And then your *active opposition*.

Some activist groups who haven't found their own frontline do one of two things. They hang out exclusively in the first wedge, building insular self-referential, marginal subcultures that are incomprehensible to everyone else. Or they behave as if *everyone* is in the last wedge, playing out the "story of the righteous few"[53] by acting like the whole world is against them. Acting from a place of interlocking frontlines means we have a broader view. When studying social movement history, we find that groups tend to win not by overpowering or beating their active opposition, but *by out-organizing them by shifting the support out from under them.*

In 1964 the Student Nonviolent Coordinating Committee (SNCC) wanted to assess the balance of forces in the United States.[54] SNCC was a major driver of the civil

52 This tool was innovated by Daniel Hunter from Training for Change.

53 See http://beyondthechoir.org/uncategorized/what-prevents-radicals-3/

54 We learned the SNCC story from Guy and Candie Carawan, eds., *Sing for Freedom: The Story of the Civil Rights Movement through Its Songs* (Bethlehem, PA: Sing Out Publications, 1990).

rights movement, and at the time they were registering black voters in the South. They understood that they didn't simply need to *educate* people, because the problem wasn't primarily ignorance or apathy. They found that they had a lot of *passive allies* who were students in the North: they were sympathetic but had no entryway into the movement. SNCC sent busses up North to bring folks down to participate in the struggle for the summer. Having a vehicle like SNCC gave the students an opportunity to align with the frontline of civil rights and engage in this historic national flashpoint. It was called Freedom Summer. Students came down in droves and for the first time witnessed lynching, violent police abuse, and angry white mobs—all simply for trying to vote. They wrote letters home to Mom and Dad, who suddenly had a personal connection to the struggle. So another shift happened: their *parents* became *passive allies*. And they brought their workplace and social networks with them. More frontlines began to intersect. Students went back to school in the fall and proceeded to organize their campuses. More shifts. The landscape in the U.S. had changed.

The cascading waterfall of support wasn't spontaneous; it was part of a planned movement strategy.

ACTION STRATEGY

> If you don't have strategy, you're part of someone else's strategy
>
> —Alvin Toffler

Within a shared social movement strategy, our organizations can find useful places to intervene with campaigns to pressure power holders for a demand. A *campaign* is a specific fight where we try to win certain demands from a *target*. A *campaign strategy* is our plan to get from point A (where we're at now) to our *goal*. Our goal is the thing we think *we* can achieve to solve the problem. The best goals are "SMART"—Specific, Measurable, Achievable, Realistic, and Time-bound—and are strategic if they are stepping stones to manifesting our *vision*. If they're not SMART, we don't have a basis to evaluate our progress and therefore can't make a plan. If our goals don't help bring about our vision then we will always be trapped in the status quo. Our vision is the way we think the world should be. Visions are big-picture, transformative, compelling, and deep. *Tactics* are the stepping stones along that plan. Tactics can range from educational measures like teach-ins, to confrontational actions like a sit-in. Our tactics escalate, and most campaigns have short and mid-term goals that build up to larger goals.

There are lots of different kinds of tactics. The broadest categories are *instrumental* and *expressive* tactics.

A tactic is *instrumental* to the degree that there is a specific quantifiable objective you are trying to achieve with it. For example, maybe we want to blockade a port that is shipping out weapons to kill people in the Middle East. We have a specific economic impact on our target, and a way to evaluate success.

A tactic is *expressive* insomuch as it expresses one's worldview, values, and identity. A mass march in response to an injustice can fall into this category. It may be useful for exciting our base, building networks and capacity, or creating a media spectacle, but usually do not have a concrete SMART goal that we can point to and say, "We achieved this specific change as a result of this tactic."

A tactic can be both instrumental and expressive. Both can be important elements. But if there is no instrumental quality to a tactic, then there are no criteria by which to evaluate our efforts.[55]

An important tactic and strategic approach of the Climate Justice movement has been building campaigns with direct action, which can be either instrumental or expressive, depending on how you use it.

The Ruckus Society defines direct action like this: *Direct Action is the strategic use of immediately effective acts to achieve a political or social end and challenge an unjust power dynamic.*

Types of Direct Action:[56]

- **Protest:** registering your dissent (rallies, marches, letter writing, petitions, e-mails, postcards, street theatre)
- **Noncooperation:** withdrawing something from the system that makes it difficult to function (boycotts, strikes, tax resistance)
- **Intervention:** directly intervening in the functioning of the system (blockading roads or buildings, disrupting meetings or "business as usual")
- **Creative Solutions:** developing alternative, community-based systems that challenge an injustice (community gardens, Food Not Bombs, off-grid housing)

While any direct action tactic requires thoughtful planning[57] (in the Ruckus Society we have about twenty-five steps that we usually use), there are several considerations in action strategy that can help with the process of aligning frontlines. If we're engaging in solidarity actions, there are several steps to take to ensure proper consultation from the frontline community we're supporting. Below is our adaptation of the Tactic Star tool originally developed by Beyond the Choir. This version of the tool offers us questions to pursue the principles of accountability outlined in the previous chapter.

The tactic star (following page) brings it all home for us, because it offers your group elicitive questions to think about the challenges around care and support, strategy, taking direction, accountability, solidarity, and trust that are discussed throughout the booklet. Often, actions are high-stress situations where the challenges of our day-to-day organizing blow up in our faces. They can be such crescendos that they highlight the need for care and intention. Use this star as a planning and preparation tool for your actions!

---

55 See http://beyondthechoir.org/uncategorized/what-prevents-radicals-1/

56 See the Ruckus Society's action manuals for more info, http://www.ruckus.org/section.php?id=17

57 For action-planning resources see, http://joshuakahnrussell.wordpress.com/resources-for-activists-and-organizers/

DEBRIEF

- What were the action's highlights? Where was there room for improvement?
- How did the action impact the base of people participating in the campaign? How did the planning deepen or strain relationships between organizers? Between groups or organizations? How did it pressure your target? What are your indicators of success?
- Don't forget to figure out next steps for follow up!

CELEBRATE

- Celebrating victories (or successful collaborations even if you haven't "won" yet) is a key piece of building relationships across difference. Celebrate now, celebrate often. Name the successes in your work, as small as they seem, it will keep the momentum going!

RESOURCES

- Is this action worth the limited time, energy, and resources of our group?
- If we are working in coalition, where does the action capacity come from? How do we distribute labor equitably?

TIMING

- Can we leverage unfolding events or new developments as opportunities?
- Does the political moment hold potential for us or vulnerability for our opponents?

TONE

- Will the action be solemn, jubilant, angry, or calm?
- Will the energy attract or repel people we want to engage?

MESSAGE

- How do you make complicated issues understandable?
- What will the tactic communicate to our audience, to our target, or to the community?
- How will the tactic carry a persuasive story?

• BEFORE YOU START: How does this tactic fit into your broader strategy and campaign plan?
• AFTER YOU FINISH: How do the lessons learned from debriefing the action reshape your campaign plan or make visible assumptions you've had in your strategy?

STRATEGY

HISTORY

• What has your group or other groups done before to address the issue?
• How have activist groups taking action in the past affected the communities living near your action location?
• What have the organizational or coalition relationships been like in the past?

COLLABORATORS

• What relationships need to be developed in order to create a collaborative direct action?
• Is the imperative for the action coming from the directly affected community themselves? If not, how do you plan to propose it?
• How will the affected constituency be involved in decision-making?

ALLIES

• How will the tactic affect your allies (the ones not working on this particular action) or potential allies?
• How does it affect community stakeholders? How will they receive it?
• Will it strengthen your relationship or jeopardize it?

AUDIENCE

• Who do we want to reach with our tactic?
• What response do we want to inspire with them?

TARGET

• What message will the tactic send to the people who have power to meet our demands?
• Will it pressure them to capitulate? Or enable them to dismiss us or retaliate?
• How will you know if you have impacted your target?

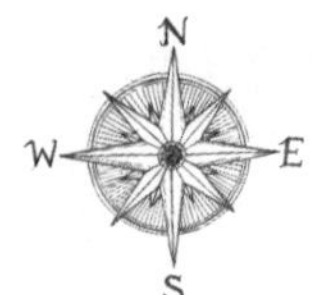

Tuning Our Political Compass

- What kinds of tactics does your organization use? Do you use them because they're the best tool for the job, or because they're what you've always done?
- How do we evaluate whether you're letting your strategy drive our tactics, instead of your tactics driving your strategy?
- How is direct action romanticized within justice movements? How might it lend itself to unaccountable or destructive behavior? How might you use it as a tool to *increase* accountability?
- How does building out your toolbox of tactics offer you more choices in practicing the pathway of accountability?

For that reason, we'd like to close the section offering one way that the Ruckus Society has gone about institutionalizing these principles and tuning its own political compass. There is no single formula to follow, but these are *excerpts of an in-process roadmap* we're developing at Ruckus. This, of course, is geared specifically for Ruckus's niche, which is nonviolent direct action. These same principles applied to your organization's niche may manifest differently. This Action Framework acts as a set of guiding principles that inform when and how we take action in mindful, accountable, and effective way.

RUCKUS ACTION FRAMEWORK

Who We Are

Ruckus supports frontline, mass-movement and people-power organizing committed to composting the current economy[58] and growing local economies based on equity, justice, direct democracy, and ecological resiliency.

Our Theory of Change

Direct Action is absolutely necessary for deep social transformation; however it must be vision-based, confrontational, honor frontline leadership, and be inclusive. We believe that bottom-up organizing and movement-building are central strategies.

Our organizing and movements must:

- **Pick a Fight:** Directly confront injustice.
- **Break the Rules:** Because the rules serve the rulers, and they keep changing the rules.
- **Build the New:** We must build our own organizations, community institutions, and projects, and we must hold them up as experiments in a better way to live together.
- **Change the Game:** Our organizing, campaigns, actions, and institutions must build power towards a transformation of our economy and an end to empire.

58 In "the current economy," we include war and empire, corporate rule, and corrupt political process.

We need direct action led by and accountable to frontline communities who are connecting the concrete improvements in their daily lives that they are fighting for with a larger transformative vision. This transformative vision is what "frames" our shared struggle and allows us to build a trans-local[59] alliance of Peoples' Movements. This movement is inclusive, inviting, inspiring, innovative, and, of course, invincible.

Our Role

To cultivate a humble and fierce Direct Action Community of Practice committed to disobedience and noncooperation with the forces of injustice. We share and expand skills and experience, inspire innovation in the movement and provide strategic movement support.

How We Work (Guiding Principles)

- **Frontline leadership:** Ruckus supports action in which frontline communities are the leader of the strategy, vision, and action.
- **Relationships for the long haul:** Ruckus comes where we're called, respecting local work and building long-term relationships of support. We believe that the first rule of accountability to communities is transparency about who we are and what we believe. We openly share our politics and vision, what we can and cannot offer, and are committed to finding the best way forward together.
- **Visionary and confrontational:** Ruckus supports action that builds strength and holds space for a strong community vision. We aspire to actions in which the visions and solutions presented are deeper and more compelling than the injustice and are placed directly in the path of the injustice.
- **Action-reflection cycle:** Ruckus is committed to ongoing organizational evolution to better represent our values, serve the movement, and respond to changing conditions in the world. We believe that our organization can itself be a living model of our vision, based on frontline and diverse leadership, shared wisdom, and the central ethic of caring and well-being for all.

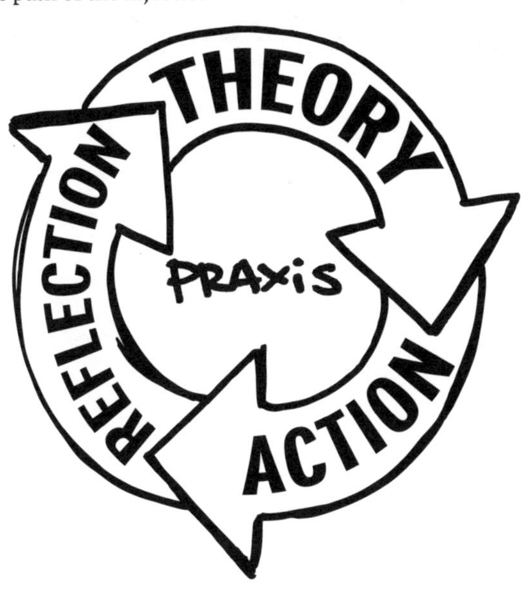

59 Trans-local means locally autonomous groups share frames, common strategies, deep solidarity, and are mass-based.

We both hope that by sharing some of the ways our main political vehicles (MCJW for Hilary; Ruckus for Joshua) are grappling with these issues they can provide a sense of how your organization, despite differences in mission, niche, tactical toolbox, arena of engagement, or strategic orientation, can begin to tune its political compass as well. We invite you into our community to practice and contribute to these ideas. Email us at organizingcoolstheplanet@gmail.com and keep up with us at www.praxismakesperfect.org and our author pages at www.pmpress.org.

JOIN THE CONVERSATION

ORGANIZINGCOOLSTHEPLANET.ORG

CHAPTER FIVE CONCLUSION

Well that was overwhelming. Writing all this felt as personal as it did analytical. Some of these lessons were hard to learn. I'm glad we kept it shorter in booklet format, but still feeling hungry to keep digging at this stuff. How are you feeling Hilary?

The same. I mean each section we have could be deepened into its own book, we tried to cover a lot...but I think it's a useful step for these conversations. We've tried to pose some key tensions and questions. I really think that just the process of trying to put ink to the page has pushed us in our own thinking and relationship to this work.

I feel like I've grown a lot through the writing process. I can't wait to see others do the same. I'm comforted by the fact that all these political frameworks are fluid. I fully hope and expect that these ideas will be replaced by the next evolution of grappling with these issues.

I'm excited that we were able to ground this stuff in our experiences. But I'm even more excited to see what kinds of conversations this generates, and to see how readers to contribute these ideas.

Yeah! If you're reading this, please email us at organizingcoolstheplanet@gmail.com with your thoughts, or add them to www.organizingcoolstheplanet.org. We'd love to hear what came up for you in reading this booklet, and whether these ideas are useful in your context. Also, you can find these and more organizing tools at www.praxismakesperfect.org. Thanks for reading!

METHODOLOGY

This booklet emerged from a variety of intentions and processes. In an effort to model what an accountability process can look like in writing a booklet like this, we engaged in a deep and extensive feedback and consultation process with the people thanked in the beginning of this booklet. We feel so grateful for their time and contributions, and feel to some degree as if we were more co*facilitators* than coauthors of this content. Additionally, these ideas came from a much broader multisectoral community of practice and we couldn't begin to list everyone who has helped shape our thinking. You know who you are. *Thank you.* Below we've outlined just a few ways this booklet came to be in our immediate processes.

- Multiple and sporadic conversations between Joshua and Hilary about these topics and shared organizing experiences.
- Previous writing each of them had done, including a zine Hilary wrote about accountability and relationship-based organizing, growing into this bigger project; and workshop curriculum from Joshua around strategy, organizing, and movement theory.
- We contacted over sixty people during the arc of the writing process for insights, direction, critiques, and feedback. This group was a mixture of frontline relationships we feel accountable to, peers, mentors, and people who fill a range of climate justice organizer roles and perspectives.
- We had several extensive one-on-one conversations with key leaders in this work to talk through the core ideas.
- Our work was inspired by and building with key concepts innovated by Movement Generation, which is more than an organization—a political community we belong to and learn from.
- Hilary completed a graduate thesis on related topics, and the interviews conducted during that project translated well, with permission, into direct quotes for this project.
- This project is open-ended, meaning that we encourage people to engage and build from these ideas to contribute to another booklet, or book.

RESOURCES

> Sit down and read. Educate yourself for the coming conflicts.
>
> —Mother Jones

Organizing and Training Toolkits:

- **Praxis Makes Perfect training resources**, http://joshuakahnrussell.wordpress.com/resources-for-activists-and-organizers/
- **Ruckus Society action planning manuals**, http://www.ruckus.org/section.php?id=82
- **New Organizing Institute Toolbox**, http://neworganizing.com/toolbox/
- **School of Unity and Liberation (SOUL),** *Youth Organizing for Community Power* **manual**, http://www.schoolofunityandliberation.org/soul_sec/resources/re-shp_manuals.html
- **smartMeme,** *Re:Imagining Change: How to Use Story-based Strategy to Win Campaigns, Build Movements, and Change the World*, http://www.smartmeme.org/downloads/smartMeme.ReImaginingChange.pdf

- **Movement Generation,** *Ecological Justice: A Call to Action,* http://www.movementgeneration.org/wp-content/uploads/2009/07/MG.CurriculumManual.pdf
- **Progressive Communicators Network Toolbox,** http://www.progressivecommunicators.net/en/toolbox
- **Rainforest Action Network, "Get Some Action: Taking Our Place in the History of U.S. Social Movements,"** http://ran.org/fileadmin/materials/comms/mediacontent/brochures/get_some_action.pdf
- **Indigenous Environmental Network, "Indigenous Peoples' Guide: False Solutions to Climate Change,"** http://www.ienearth.org/docs/Indigenous_Peoples_Guide-E.pdf

Statements, Declarations, and Frameworks:

- **Jemez Principles for Democratic Organizing,** http://www.ejnet.org/ej/jemez.pdf
- **Bali Principles of Climate Justice,** http://www.cbecal.org/pdf/bali-principles.pdf
- **CJN!NA Manifesto,** http://joshuakahnrussell.files.wordpress.com/2008/10/cjn-na-manifesto.doc
- **CJ IN THE USA: Root Cause Remedies, Rights, Reparations, and Representation,** http://www.climate-justice-now.org/cj-in-the-usa-root-cause-remedies-rights-reparations-and-representation/
- **US Social Forum EcoJustice Peoples' Movement Assembly Declaration,** http://joshuakahnrussell.files.wordpress.com/2008/10/ecojustice-resolution.doc
- **Indigenous Environmental Network's Four Climate Justice Principles,** http://www.ienearth.org/docs/IEN_4_Principles_of_Climate_Justice.pdf

Organizations Referenced in This Booklet:

- **Indigenous Environmental Network,** www.ienearth.org
- **Mobilization for Climate Justice West,** west.actforclimatejustice.org
- **Ruckus Society,** www.ruckus.org
- **smartMeme,** www.smartmeme.org
- **Beyond the Choir,** www.beyondthechoir.org
- **Climate Justice Now!,** www.climate-justice-now.org
- **Institute for Policy Studies,** www.ips-dc.org
- **Movement Generation,** www.movementgeneration.org

ABOUT THE AUTHORS

Joshua Kahn is a strategy, organizing, and nonviolent direct action trainer with the Ruckus Society. He has worked internationally with the Climate Justice Now! network to bring justice to the United Nations Framework Convention on Climate Change and has been a leading voice within the International Youth Climate Movement. Joshua spent four years as Rainforest Action Network's grassroots actions manager, campaigning against banks and corporations to end our addiction to coal and oil. He has authored chapters for numerous books, most recently *The Next Eco-Warriors*. His articles have appeared in *Yes!* magazine, *Left Turn*, *PeaceWork* magazine, *Upping the Anti*, and *Z Magazine*. His blog is www.praxismakesperfect.org.

Hilary Moore has been organizing around social justice issues in the Bay Area for the past four years. She is a core organizer with Rising Tide Bay Area and a founding organizer of the Mobilization for Climate Justice West, a grassroots alliance of organizations in the Bay Area dedicated to keeping frontline communities at the forefront of the struggle while advancing community-led solutions. Hilary organizes around gentrification issues in West Oakland, as well as organizing allied support for the California Domestic Workers Bill of Rights. She sits on the board of the Institute for Social Ecology. Her interests and research focus on building collective practices of care within communities engaged in resistance and struggle.

PM Press is an independent, radical publisher of books and media to educate, entertain, and inspire. Founded in 2007 by a small group of people with decades of publishing, media, and organizing experience, PM Press amplifies the voices of radical authors, artists, and activists. Our aim is to deliver bold political ideas and vital stories to all walks of life and arm the dreamers to demand the impossible. We have sold millions of copies of our books, most often one at a time, face to face. We're old enough to know what we're doing and young enough to know what's at stake. Join us to create a better world.

PM Press
PO Box 23912
Oakland, CA 94623
510-658-3906 • info@pmpress.org

PM Press in Europe
europe@pmpress.org
www.pmpress.org.uk

FRIENDS OF PM

These are indisputably momentous times—the financial system is melting down globally and the Empire is stumbling. Now more than ever there is a vital need for radical ideas.

In the many years since its founding—and on a mere shoestring—PM Press has risen to the formidable challenge of publishing and distributing knowledge and entertainment for the struggles ahead. With hundreds of releases to date, we have published an impressive and stimulating array of literature, art, music, politics, and culture. Using every available medium, we've succeeded in connecting those hungry for ideas and information to those putting them into practice.

Friends of PM allows you to directly help impact, amplify, and revitalize the discourse and actions of radical writers, filmmakers, and artists. It provides us with a stable foundation from which we can build upon our early successes and provides a much-needed subsidy for the materials that can't necessarily pay their own way. You can help make that happen—and receive every new title automatically delivered to your door once a month—by joining as a Friend of PM Press. And, we'll throw in a free T-shirt when you sign up.

Here are your options:

- $30 a month: Get all books and pamphlets plus 50% discount on all webstore purchases
- $40 a month: Get all PM Press releases (including CDs and DVDs) plus 50% discount on all webstore purchases
- $100 a month: Superstar—Everything plus PM merchandise, free downloads, and 50% discount on all webstore purchases

For those who can't afford $30 or more a month, we have Sustainer Rates at $15, $10, and $5. Sustainers get a free PM Press T-shirt and a 50% discount on all purchases from our website.

Your Visa or Mastercard will be billed once a month, until you tell us to stop. Or until our efforts succeed in bringing the revolution around. Or the financial meltdown of Capital makes plastic redundant. Whichever comes first.

Re:Imagining Change

How to Use Story-based Strategy to Win Campaigns, Build Movements, and Change the World Second Edition

Patrick Reinsborough and Doyle Canning
Foreword by Jonathan Matthew Smucker
978-1-62963-384-8 • $18.95

Re:Imagining Change provides resources, theory, hands-on tools, and illuminating case studies for the next generation of innovative change-makers. This unique book explores how culture, media, memes, and narrative intertwine with social change strategies, and offers practical methods to amplify progressive causes in the popular culture.

Re:Imagining Change is an inspirational inside look at the trailblazing methodology developed by the Center for Story-based Strategy over fifteen years of their movement building partnerships. This practitioner's guide is an impassioned call to innovate our strategies for confronting the escalating social and ecological crises of the twenty-first century. This new, expanded second edition includes updated examples from the frontlines of social movements and provides the reader with easy-to-use tools to change the stories they care about most.

> "We are surrounded and shaped by stories every day, sometimes for better, sometimes for worse. But what Doyle Canning and Patrick Reinsborough point out is a beautiful and powerful truth—that we are all storytellers too. Armed with the right narrative tools, activists can not only open the world's eyes to injustice, but feed the desire for a better world. *Re:Imagining Change* is a powerful weapon for a more democratic, creative and hopeful future."
> —Raj Patel, author of *The Value of Nothing: How to Reshape Market Society and Redefine Democracy*

Praise for *Organizing Cools the Planet*

"As the climate crisis becomes increasingly unignorable, our movements must learn to navigate a rapidly changing and high-stakes political landscape. Our times demand we think bigger, push harder, and reimagine the possibilities for twenty-first-century movement building. This potent booklet is a great place to begin the conversation. Authored by two visionary young leaders who share their personal struggles and hard-earned lessons from organizing at the intersection of justice, ecology, and change, *Organizing Cools the Planet* is required reading for anyone who gives a damn about the future. Tune in for some indispensable analysis, provocative thinking and a healthy dose of people-powered optimism."

—Patrick Reinsborough, cofounder,
smartMeme Strategy & Training Project

"This is a rigorous and useful tool for teaching and learning the architecture of organizing, a valuable nourishment for climate justice activists and change agents."

—Dorothy Guerrero,
Focus on the Global South

"It is an erudite manual, spirited and consistently engaging."

—Andrej Grubačić, author, *Don't Mourn, Balkanize!*
and *Wobblies and Zapatistas*

"Still young and developing, the climate justice movement has already shaken up politics with its holistic perspective and fresh energy. *Organizing Cools the Planet* offers a set of tools to help this dynamic new movement sharpen its strategies, promote frontline leadership, and realize its tremendous potential."

—Max Elbaum, cofounder,
WarTimes/Tiempo de Guerras; author